Elementary Applications of Probability Theory

CHAPMAN & HALL STATISTICS TEXTBOOK SERIES

Editors:

Dr Chris Chatfield
Reader in Statistics
School of Mathematical Sciences
University of Bath, UK

Professor Jim V. Zidek
Department of Statistics
University of British Columbia, Canada

OTHER TITLES IN THE SERIES INCLUDE

Practical Statistics for Medical Research
D.G. Altman

Interpreting Data
A.J.B. Anderson

Statistical Methods for SPC and TQM
D. Bissell

Statistics in Research and Development
Second edition
R. Caulcutt

The Analysis of Time Series
Fourth edition
C. Chatfield

Problem Solving – A Statistician's Guide
C. Chatfield

Statistics for Technology
Third edition
C. Chatfield

Introduction to Multivariate Analysis
C. Chatfield and A.J. Collins

Modelling Binary Data
D. Collett

Modelling Survival Data in Medical Research
D. Collett

Applied Statistics
D.R. Cox and E.J. Snell

Statistical Analysis of Reliability Data
M.J. Crowder, A.C. Kimber, T.J. Sweeting and R.L. Smith

An Introduction to Generalized Linear Models
A.J. Dobson

Multivariate Analysis of Variance and Repeated Measures
D.J. Hand and C.C. Taylor

The Theory of Linear Models
B. Jorgensen

Statistical Theory
Fourth edition
B. Lindgren

Essential Statistics
Second edition
D.G. Rees

Decision Analysis: A Bayesian Approach
J.Q. Smith

Applied Nonparametric Statistical Methods
Second edition
P. Sprent

Elementary Applications of Probability Theory
H.C. Tuckwell

Statistical Process Control: Theory and Practice
Third edition
G.B. Wetherill and D.W. Brown

Statistics in Engineering
A practical approach
A.V. Metcalfe

Full information on the complete range of Chapman & Hall statistics books is available from the publishers.

Elementary Applications of Probability Theory

With an introduction to stochastic differential equations

Second edition

Henry C. Tuckwell

Senior Research Fellow
Stochastic Analysis Group of the
Centre for Mathematics and its Applications
Australian National University
Australia

CRC Press
Taylor & Francis Group
Boca Raton London New York

CRC Press is an imprint of the
Taylor & Francis Group, an **informa** business

A CHAPMAN & HALL BOOK

First published 1988 by Chapman & Hall

First edition 1988
Second edition 1995

Published 2019 by CRC Press
Taylor & Francis Group
6000 Broken Sound Parkway NW, Suite 300
Boca Raton, FL 33487-2742

© 1988, 1995 Henry C. Tuckwell
CRC Press is an imprint of Taylor & Francis Group, an Informa business

First issued in paperback 2019

No claim to original U.S. Government works

ISBN-13: 978-0-367-44905-6 (pbk)
ISBN-13: 978-0-412-57620-1 (hbk)

This book contains information obtained from authentic and highly regarded sources. Reasonable efforts have been made to publish reliable data and information, but the author and publisher cannot assume responsibility for the validity of all materials or the consequences of their use. The authors and publishers have attempted to trace the copyright holders of all material reproduced in this publication and apologize to copyright holders if permission to publish in this form has not been obtained. If any copyright material has not been acknowledged please write and let us know so we may rectify in any future reprint.

Except as permitted under U.S. Copyright Law, no part of this book may be reprinted, reproduced, transmitted, or utilized in any form by any electronic, mechanical, or other means, now known or hereafter invented, including photocopying, microfilming, and recording, or in any information storage or retrieval system, without written permission from the publishers.

Trademark Notice: Product or corporate names may be trademarks or registered trademarks, and are used only for identification and explanation without intent to infringe.

Visit the Taylor & Francis Web site at
http://www.taylorandfrancis.com

and the CRC Press Web site at
http://www.crcpress.com

Typeset in 10/12 pt Times by Thomson Press (India) Ltd, New Delhi

A catalogue record for this book is available from the British Library

Library of Congress Catalog Card Number: 94-68995

To Silvia Dori

To Silvia, Don

Contents

x   Contents

Preface

For the second edition, two additional chapters, Chapters 11 and 12, have been written. The added material should make the book suitable for two consecutive courses in elementary and intermediate applications of probability.

The new material consists of an introduction to stochastic differential equations. It is hoped that this will be useful for applied mathematical modelling of the behaviour of many naturally occurring randomly fluctuating quantities. An attempt has been made to explain the material with a certain amount of rigour, but hopefully without so much detail that a practical understanding is impaired.

The stochastic differential equations in this book are first order equations with an additional noise term. This added term usually contains a Gaussian 'white noise' so that the resulting solution is called a *diffusion process*.

Chapter 11 starts with a brief reminder of the nature of ordinary deterministic differential equations, followed by an explanation of the essential differences between deterministic and stochastic equations. These have been illustrated with data in neurophysiology and economics.

There follows a thorough discussion of the properties of the standard Wiener process which forms a cornerstone of the theory, and a section on white noise which is a useful concept, especially for modelling. The simplest stochastic differential equations, being those of the Wiener process with drift, are then introduced.

The analytical approach, by which is meant a study of such quantities as transition probabilities through the equations they satisfy, is introduced at the end of Chapter 11. This method of study is essential if one is to obtain accurate estimates for many properties of the random processes one is considering.

Chapter 12 starts with a continuation of the analytic method by introducing the equations of Kolmogorov (or Fokker-Planck). Steady state and time-dependent solutions of these equations are then found for problems of classical interest - those involving the Wiener and Ornstein-Uhlenbeck processes.

In section 12.5, Ito's stochastic integral and the corresponding stochastic differential equations are introduced. This is followed by a heuristic derivation of the formula for change of variable. A brief treatment is given of Stratonovich's integral which leads to simpler rules of calculus.

Sections 12.6 and 12.7 concern certain aspects of modelling with stochastic differential equations. The connection between the analytical method and the *direct* method involving stochastic differential equations is given, followed by examples drawn from theoretical population biology, mathematical economics and theoretical neurobiology.

Finally, a brief discussion is undertaken of various practical considerations–solution of stochastic differential equations by simulation, numerical methods of solution of equations for transition probabilities, and estimation of parameters.

Both Chapters 11 and 12 end with sets of exercises, some of those in Chapter 12 being suitable for project work.

It is a pleasure to thank the following colleagues for helpful suggestions with respect to the new material: Professors Joseph Gani and Christopher C. Heyde of the Institute of Advanced Studies in Canberra; and Professor James A. Koziol of Scripps Research Institute, La Jolla.

Henry C. Tuckwell
Versailles, June 1994

Preface to the first edition

This book concerns applications of probability theory. It has been written in the hope that the techniques presented will be useful for problems in diverse areas. A majority of the examples come from the biological sciences but the concepts and techniques employed are not limited to that field. To illustrate, birth and death processes (Chapter 9) have applications to chemical reactions, and branching processes (Chapter 10) have applications in physics but neither of these specific applications is developed in the text.

The book is based on an undergraduate course taught to students who have had one introductory course in probability and statistics. Hence it does not contain a lengthy introduction to probability and random variables, for which there are many excellent books. Prerequisites also include an elementary knowledge of calculus, including first-order differential equations, and linear algebra.

The basic plan of the book is as follows.

Chapter 1: a review of basic probability theory;
Chapters 2-5: random variables and their applications;
Chapter 6: sequences of random variables and concepts of convergence;
Chapters 7-10: theory and properties of basic random processes.

The outline is now given in more detail.

Chapter 1 contains a brief review of some of the basic material which will be needed in later chapters; for example, the basic probability laws, conditional probability, change of variables, etc. It is intended that Chapter 1 be used as a reference rather than a basis for instruction. Students might be advised to study this chapter as the material is called upon.

Chapter 2 illustrates the interplay between geometry and probability. It begins with an historically interesting problem and then addresses the problem of finding the density of the distance between two randomly chosen points. The second such case, when the points occur within a circle, is not easy but the result is useful.

Chapter 3 begins with the properties of the hypergeometric distribution. An important application is developed, namely the estimation of animal populations by the capture–recapture method. The Poisson distribution is then reviewed and one-dimensional Poisson point processes introduced together with some of their basic properties. There follows a generalization to two dimensions, which enables one to study spatial distributions of plants and to develop methods to estimate their population numbers. The chapter concludes with the compound Poisson distribution which is illustrated by application to a neurophysiological model.

Chapter 4 introduces several of the basic concepts of reliability theory. The relevant properties of the standard failure time distributions are given. The interesting spare parts problem is next and the concluding sections discuss methods for determining the reliability of complex systems.

Chapter 5 commences by explaining the usefulness of computer simulation. There follows an outline of the theory of random number generation using the linear congruential method and the probability integral transformation. The polar method for normal random variables is given. Finally, tests for the distribution and independence properties of random numbers are described.

Chapter 6 deals with sequences of random variables. Some methods for studying convergence in distribution and convergence in probability are developed. In particular, characteristic functions and Chebyshev's inequality are the main tools invoked. The principal applications are to proving a central limit theorem and a weak law of large numbers. Several uses for the latter are detailed.

Chapter 7 starts with the definition of random (stochastic) processes and introduces the important Markov property. The rest of the chapter is mainly concerned with the elementary properties of simple random walks. Included are the unrestricted process and that in the presence of absorbing barriers. For the latter the probability of absorption and the expected time of absorption are determined using the difference equation approach. The concluding section briefly introduces the Wiener process, so fundamental in advanced probability. The concept of martingale and its usefulness are discussed in the exercises.

Chapter 8 is on Markov chains. However, the theory is motivated by examples in population genetics, so the Hardy-Weinberg principle is discussed first. Elementary general Markov chain theory is developed for absorbing Markov chains and those with stationary distributions.

Chapter 9 concerns birth and death processes, which are motivated by demographic considerations. The Poisson process is discussed as a birth process because of its fundamental role. There follow the properties of the Yule process, a simple death process and the simple birth and death process. The treatment of the latter only states rather than derives the equation satisfied by the probability generating function but this enables one to derive the satisfying result concerning the probability of extinction.

Chapter 10 contains a brief introduction to the theory of branching processes, focusing on the standard Galton-Watson process. It is motivated by the phenomenon of cell division. The mean and variance are derived and the probability of extinction determined.

It should be mentioned that references are sometimes not to the latest editions of books; for example, those of Hoel, Pielou, Strickberger and Watson.

In the author's view there is ample material for a one-quarter or one-semester course. In fact some material might have to be omitted in such a course. Alternatively, the material could be presented in two courses, with a division at Chapter 6, supplemented by further reading in specialist areas (e.g. ecology, genetics, reliability, psychology) and project work (e.g. simulation).

I thank the many Monash students who have taken the course in applied probability on which this book is based. In particular, Derryn Griffiths made many useful suggestions. It is also a pleasure to acknowledge the helpful criticisms of Dr James A. Koziol of Scripps Clinic and Research Foundation, La Jolla; and Drs Fima Klebaner and Geoffrey A. Watterson at Monash University. I am also grateful to Barbara Young for her excellent typing and to Jean Sheldon for her splendid artwork.

Henry C. Tuckwell
Los Angeles, April 1987

1
A review of basic probability theory

This is a book about the applications of probability. It is hoped to convey that this subject is both a fascinating and important one. The examples are drawn mainly from the biological sciences but some originate in the engineering, physical, social and statistical sciences. Furthermore, the techniques are not limited to any one area.

The reader is assumed to be familiar with the elements of probability or to be studying it concomitantly. In this chapter we will briefly review some of this basic material. This will establish notation and provide a convenient reference place for some formulas and theorems which are needed later at various points.

1.1 PROBABILITY AND RANDOM VARIABLES

When an experiment is performed whose outcome is uncertain, the collection of possible **elementary outcomes** is called a **sample space**, often denoted by Ω. Points in Ω, denoted in the discrete case by ω_i, $i = 1, 2, \ldots$ have an associated probability $P\{\omega_i\}$. This enables the probability of any subset A of Ω, called an **event**, to be ascertained by finding the total probability associated with all the points in the given subset:

$$P\{A\} = \sum_{\omega_i \in A} P\{\omega_i\}$$

We always have

$$0 \leqslant P\{A\} \leqslant 1,$$

and in particular $P\{\Omega\} = 1$ and $P\{\varnothing\} = 0$, where $\varnothing$ is the empty set relative to Ω.

A **random variable** is a real-valued function defined on the elements of a sample space. Roughly speaking it is an observable which takes on numerical values with certain probabilities.

Discrete random variables take on finitely many or countably infinitely many values. Their probability laws are often called **probability mass functions**. The following discrete random variables are frequently encountered.

Binomial

A **binomial** random variable X with parameters n and p has the probability law

$$p_k = \Pr\{X = k\} = \binom{n}{k} p^k q^{n-k} \tag{1.1}$$

$$\doteq b(k; n, p), \qquad k = 0, 1, 2, \ldots, n,$$

where $0 \leqslant p \leqslant 1$, $q = 1 - p$ and n is a positive integer ($\doteq$ means we are defining a new symbol). The **binomial coefficients** are

$$\binom{n}{k} = \frac{n!}{k!(n-k)!},$$

being the number of ways of choosing k items, without regard for order, from n distinguishable items.

When $n = 1$, so we have

$$\Pr\{X = 1\} = p = 1 - \Pr\{X = 0\},$$

the random variable is called **Bernoulli**.

Note the following.

Convention

Random variables are always designated by capital letters (e.g. X, Y) whereas symbols for the values they take on, as in $\Pr\{X = k\}$, are always designated by lowercase letters.

The converse, however, is not true. Sometimes we use capital letters for non-random quantities.

Poisson

A **Poisson** random variable with parameter $\lambda > 0$ takes on non-negative integer values and has the probability law

$$p_k = \Pr\{X = k\} = \frac{e^{-\lambda}\lambda^k}{k!}, \qquad k = 0, 1, 2, \ldots. \tag{1.2}$$

For any random variable the total probability mass is unity. Hence if p_k is given by either (1.1) or (1.2),

$$\sum_k p_k = 1$$

where summation is over the possible values k as indicated.

For any random variable X, the **distribution function** is

$$F(x) = \Pr\{X \leqslant x\}, \quad -\infty < x < \infty.$$

Continuous random variables take on a continuum of values. Usually the probability law of a continuous random variable can be expressed through its **probability density function,** $f(x)$, which is the derivative of the distribution function. Thus

$$f(x) = \frac{d}{dx} F(x)$$

$$= \lim_{\Delta x \to 0} \frac{F(x + \Delta x) - F(x)}{\Delta x}$$

$$= \lim_{\Delta x \to 0} \frac{\Pr\{X \leqslant x + \Delta x\} - \Pr\{X \leqslant x\}}{\Delta x} \qquad (1.3)$$

$$= \lim_{\Delta x \to 0} \frac{\Pr\{x < X \leqslant x + \Delta x\}}{\Delta x}$$

$$= \lim_{\Delta x \to 0} \frac{\Pr\{X \in (x, x + \Delta x]\}}{\Delta x}$$

The last two expressions in (1.3) often provide a convenient prescription for calculating probability density functions. Often the latter is abbreviated to p.d.f. but we will usually just say 'density'.

If the interval (x_1, x_2) is in the range of X then the probability that X takes values in this interval is obtained by integrating the probability density over (x_1, x_2).

$$\Pr\{x_1 < X < x_2\} = \int_{x_1}^{x_2} f(x)\,dx.$$

The following continuous random variables are frequently encountered.

Normal (or Gaussian)

A random variable with density

$$f(x) = \frac{1}{\sqrt{2\pi\sigma^2}} \exp\left\{ -\frac{(x - \mu)^2}{2\sigma^2} \right\}, \quad -\infty < x < \infty, \qquad (1.4)$$

where $-\infty < \mu < \infty$ and $0 < \sigma^2 < \infty$,

is called **normal**. The quantities μ and σ^2 are the mean and variance (elaborated upon below) and such a random variable is often designated

$N(\mu, \sigma)$. If $\mu = 0$ and $\sigma = 1$ the random variable is called a **standard normal** random variable, for which the usual symbol is Z.

Uniform

A random variable with constant density

$$f(x) = \frac{1}{b-a}, \qquad -\infty < a \leqslant x \leqslant b < \infty,$$

is said to be **uniformly distributed** on (a, b) and is denoted $U(a, b)$. If $a = 0, b = 1$ the density is unity on the unit interval,

$$f(x) = 1, \qquad 0 \leqslant x \leqslant 1$$

and the random variable is designated $U(0, 1)$.

Gamma

A random variable is said to have a **gamma density** (or gamma distribution) with parameters λ and ρ if

$$f(x) = \frac{\lambda(\lambda x)^{\rho - 1} e^{-\lambda x}}{\Gamma(\rho)}, \qquad x \geqslant 0; \qquad \lambda, \rho > 0.$$

The quantity $\Gamma(\rho)$ is the **gamma function** defined as

$$\Gamma(\rho) = \int_0^\infty x^{\rho - 1} e^{-x} \, dx, \qquad \rho > 0.$$

When $\rho = 1$ the gamma density is that of an **exponentially distributed random variable**

$$f(x) = \lambda e^{-\lambda x}, \qquad x > 0.$$

For continuous random variables the density must integrate to unity:

$$\int f(x) \, dx = 1$$

where the interval of integration is the whole range of values of X.

1.2 MEAN AND VARIANCE

Let X be a discrete random variable with

$$\Pr\{X = x_k\} = p_k, \qquad k = 1, 2, \ldots.$$

The **mean, average** or **expectation** of X is

$$E(X) = \sum_k p_k x_k.$$

For a binomial random variable $E(X) = np$ whereas a Poisson random variable has mean $E(X) = \lambda$.

For a continuous random variable with density $f(x)$,

$$E(X) = \int x f(x) \, dx.$$

If X is normal with density given by (1.4) then $E(X) = \mu$; a uniform (a, b) random variable has mean $E(X) = \frac{1}{2}(a + b)$; and a gamma variate has mean $E(X) = \rho/\lambda$.

The **nth moment** of X is the expected value of X^n:

$$E(X^n) = \begin{cases} \sum_k p_k x_k^n & \text{if } X \text{ is discrete,} \\ \int x^n f(x) \, dx & \text{if } X \text{ is continuous.} \end{cases}$$

If $n = 2$ we obtain the **second moment** $E(X^2)$. The **variance**, which measures the degree of dispersion of the probability mass of a random variable about its mean, is

$$\text{Var}(X) = E[(X - E(X))^2]$$

$$= E(X^2) - E^2(X).$$

The variances of the above-mentioned random variables are:

binomial, npq; Poisson, λ; normal, σ^2; uniform, $\frac{1}{12}(b - a)^2$; gamma, ρ/λ^2.

The square root of the variance is called the **standard deviation**.

1.3 CONDITIONAL PROBABILITY AND INDEPENDENCE

Let A and B be two random events. The **conditional probability** of A given B is, provided $\Pr\{B\} \neq 0$,

$$\Pr\{A|B\} = \frac{\Pr\{AB\}}{\Pr\{B\}}$$

where AB is the **intersection** of A and B, being the event that both A and B occur (sometimes written $A \cap B$). Thus only the occurrences of A which are simultaneous with those of B are taken into account. Similarly, if X, Y are random variables defined on the same sample space, taking on values $x_i, i = 1, 2, \ldots, y_j, j = 1, 2, \ldots$, then the conditional probability that $X = x_i$ given $Y = y_j$ is, if $\Pr\{Y = y_j\} \neq 0$,

$$\Pr\{X = x_i | Y = y_j\} = \frac{\Pr\{X = x_i, Y = y_j\}}{\Pr\{Y = y_j\}},$$

the comma between $X = x_i$ and $Y = y_j$ meaning 'and'.

The **conditional expectation** of X given $Y = y_j$ is

$$E(X | Y = y_j) = \sum_i x_i \Pr\{X = x_i | Y = y_j\}.$$

The **expected value** of XY is

$$E(XY) = \sum_{i,j} x_i y_j \Pr\{X = x_i, Y = y_j\},$$

and the **covariance** of X, Y is

$$\mathrm{Cov}(X, Y) = E[(X - E(X))(Y - E(Y))]$$
$$= E(XY) - E(X)E(Y).$$

The covariance is a measure of the linear dependence of X on Y.

If X, Y are independent then the value of Y should have no effect on the probability that X takes on any of its values. Thus we define X, Y as **independent** if

$$\Pr\{X = x_i | Y = y_j\} = \Pr\{X = x_i\}, \quad \text{all } i,j.$$

Equivalently X, Y are independent if

$$\Pr\{X = x_i, Y = y_j\} = \Pr\{X = x_i\} \Pr\{Y = y_j\},$$

with a similar formula for arbitrary independent events.

Hence for independent random variables

$$E(XY) = E(X)E(Y),$$

so their covariance is zero. Note, however, that $\mathrm{Cov}(X, Y) = 0$ does not always imply X, Y are independent. The covariance is often normalized by defining the **correlation coefficient**

$$\rho_{XY} = \frac{\mathrm{Cov}(X, Y)}{\sigma_X \sigma_Y}$$

where σ_X, σ_Y are the standard deviations of X, Y. ρ_{XY} is bounded above and below by

$$\boxed{-1 \leqslant \rho_{XY} \leqslant 1}$$

Let $X_1, X_2, \ldots, X_n$ be **mutually independent** random variables. That is,

$$\Pr\{X_1 \in A_1, X_2 \in A_2, \ldots, X_n \in A_n\}$$
$$= \Pr\{X_1 \in A_1\}\,\Pr\{X_2 \in A_2\}\ldots\Pr\{X_n \in A_n\},$$

for all appropriate sets $A_1, \ldots, A_n$. Then

$$\boxed{\operatorname{Var}\left(\sum_{i=1}^{n} X_i\right) = \sum_{i=1}^{n} \operatorname{Var}(X_i)}$$

so that variances add in the case of independent random variables. We also note the formula

$$\operatorname{Var}(aX + bY) = a^2 \operatorname{Var}(X) + b^2 \operatorname{Var}(Y),$$

which holds if X, Y are independent. If $X_1, X_2, \ldots, X_n$ are **independent identically distributed** (abbreviated to **i.i.d.**) random variables with $E(X_1) = \mu$, $\operatorname{Var}(X_1) = \sigma^2$, then

$$E\left(\sum_{i=1}^{n} X_i\right) = \mu n; \qquad \operatorname{Var}\left(\sum_{i=1}^{n} X_i\right) = n\sigma^2.$$

If X is a random variable and $\{X_1, X_2, \ldots, X_n\}$ are i.i.d. with the distribution of X, then the collection $\{X_k\}$ is called a **random sample of size n** for X. Random samples play a key role in computer simulation (Chapter 5) and of course are fundamental in statistics.

1.4 LAW OF TOTAL PROBABILITY

Let Ω be a sample space for a random experiment and let $\{A_i, i = 1, 2, \ldots\}$ be a collection of nonempty subsets of Ω such that

(i) $A_i A_j = \varnothing$, $i \neq j$;
(ii) $\underset{i}{\cup}\, A_i = \Omega$.

(Here $\varnothing$ is the null set, the impossible event, being the complement of Ω.) Condition (i) says that the A_i represent **mutually exclusive** events. Condition (ii) states that when an experiment is performed, at least one of the A_i must be observed. Under these conditions the sets or events $\{A_i, i = 1, 2, \ldots\}$ are said to form a **partition** or **decomposition** of the sample space.

The **law or theorem of total probability** states that for any event (set) B,

$$\Pr\{B\} = \sum_i \Pr\{B|A_i\}\,\Pr\{A_i\}$$

A similar relation holds for expectations. By definition the expectation of X conditioned on the event A_i is

$$E(X|A_i) = \sum_k x_k \Pr\{X = x_k|A_i\},$$

where $\{x_k\}$ is the set of possible values of X. Thus

$$E(X) = \sum_k x_k \Pr\{X = x_k\}$$

$$= \sum_k x_k \sum_i \Pr\{X = x_k|A_i\}\,\Pr\{A_i\}$$

$$= \sum_i \Pr\{A_i\} \sum_k x_k \Pr\{X = x_k|A_i\}.$$

Thus

$$E(X) = \sum_i E(X|A_i)\,\Pr\{A_i\}$$

which we call the **law of total probability applied to expectations.**

We note also the fundamental relation for any two events A, B in the same sample space:

$$\Pr\{A \cup B\} = \Pr\{A\} + \Pr\{B\} - \Pr\{AB\}$$

where $A \cup B$ is the **union** of A and B, consisting of those points which are in A or in B or in both A and B.

1.5 CHANGE OF VARIABLES

Let X be a continuous random variable with distribution function F_X and density f_X. Let

$$y = g(x)$$

be a strictly increasing function of x (see Fig. 1.1) with inverse function

$$x = h(y).$$

Then

$$Y = g(X)$$

is a random variable which we let have distribution function F_Y and density f_Y.

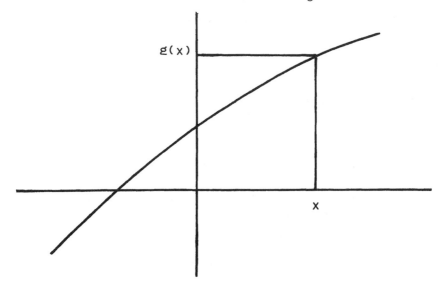

Figure 1.1 $g(x)$ is a strictly increasing function of x.

It is easy to see that $X \leqslant x$ implies $Y \leqslant g(x)$. Hence we arrive at

$$\Pr\{X \leqslant x\} = \Pr\{Y \leqslant g(x)\}$$

By the definition of distribution functions this can be written

$$F_X(x) = F_Y(g(x)). \tag{1.5}$$

Therefore

$$F_Y(y) = F_X(h(y)).$$

On differentiating with respect to y we obtain, assuming that h is differentiable,

$$\frac{dF_Y}{dy} = \frac{dF_X(x)}{dx}\bigg|_{h(y)} \frac{dh}{dy}$$

or in terms of densities

$$f_Y(y) = f_X(h(y))\frac{dh}{dy}. \tag{1.6}$$

If y is a strictly decreasing function of x we obtain

$$\Pr\{X \leqslant x\} = \Pr\{Y \geqslant g(x)\}.$$

Working through the steps between (1.5) and (1.6) in this case gives

$$f_Y(y) = f_X(h(y))\left(-\frac{dh}{dy}\right). \tag{1.7}$$

Both formulas (1.6) and (1.7) are covered by the single formula

$$f_Y(y) = f_X(h(y)) \left| \frac{dh}{dy} \right|$$

where $|\ \ |$ denotes **absolute value**. Cases where g is neither strictly increasing nor strictly decreasing require special consideration.

1.6 TWO-DIMENSIONAL RANDOM VARIABLES

Let X, Y be random variables defined on the same sample space. Then their **joint distribution function** is

$$F_{XY}(x, y) = \Pr\{X \leqslant x, Y \leqslant y\}.$$

The mixed partial derivative of F_{XY}, if it exists, is the **joint density** of X and Y:

$$f_{XY}(x, y) = \frac{\partial^2 F_{XY}}{\partial x \partial y}.$$

As a rough guide we have, for small enough $\Delta x, \Delta y$,

$$f_{XY}(x, y)\Delta x\Delta y \simeq \Pr\{X \in (x, x + \Delta x], Y \in (y, y + \Delta y]\}.$$

If X, Y are independent then their joint distribution function and joint density function factor into those of the individual random variables:

$$F_{XY}(x, y) = F_X(x)F_Y(y),$$
$$f_{XY}(x, y) = f_X(x)f_Y(y).$$

In particular, if X, Y are **independent standard normal random variables**,

$$f_{XY}(x, y) = \left(\frac{1}{\sqrt{2\pi}} \exp\left\{ \frac{-x^2}{2} \right\} \right)\left(\frac{1}{\sqrt{2\pi}} \exp\left\{ \frac{-y^2}{2} \right\} \right), \quad -\infty < x, y < \infty,$$

which can be written

$$f_{XY}(x, y) = \frac{1}{2\pi} \exp\{ - \tfrac{1}{2}(x^2 + y^2) \} \tag{1.8}$$

In fact if the joint density X, Y is as given by (1.8) we may conclude that X, Y are independent standard normal random variables.

Change of variables

Let U, V be random variables with joint density $f_{UV}(u, v)$. Suppose that the one–one mappings

$$X = G_1(U, V)$$
$$Y = G_2(U, V)$$

transform U, V to the pair of random variables X, Y. Let the inverse transformations be

$$U = H_1(X, Y)$$
$$V = H_2(X, Y).$$

Then the joint density of X, Y is given by

$$f_{XY}(x, y) = f_{UV}(H_1(x, y), H_2(x, y)) |J(x, y)|$$

where $J(x, y)$ is the **Jacobian** of the inverse transformation given by

$$J(x, y) = \begin{vmatrix} \dfrac{\partial H_1}{\partial x} & \dfrac{\partial H_1}{\partial y} \\[2mm] \dfrac{\partial H_2}{\partial x} & \dfrac{\partial H_2}{\partial y} \end{vmatrix}$$

and $|J|$ is its absolute value. A proof of this result is given in Blake (1979).

1.7 HYPOTHESIS TESTING - THE χ^2 GOODNESS OF FIT TEST

In testing the validity of a **stochastic** (also called **random, probabilistic**) **model** it is often necessary to perform statistical tests on data. The basic idea is to consider a random variable which can be observed when the random experiment of interest is performed. Such a random variable is called a **test statistic**. If under a given hypothesis values of a test statistic occur (when the experiment is performed) which are considered unlikely, one is inclined to reject that hypothesis.

χ^2 random variables

Apart from a test for independence developed in Section 5.6, the only statistical test which is used in this book is called the χ^2 **goodness of fit test**. We first define χ^2 random variables and then see how these are useful in testing hypotheses about probability distributions.

Definition X_n **is a χ^2-random variable with n degrees of freedom if its density is**

$$f_n(x) = \frac{1}{2^{n/2}\Gamma(n/2)} x^{n/2 - 1} e^{-x/2}, \qquad x > 0; \quad n = 1, 2, \ldots \qquad (1.9)$$

The mean and variance of such a random variable are given by

$$E(X_n) = n,$$
$$\text{Var}(X_n) = 2n.$$

Also, it may be shown that the density (1.9) is that of a sum of squares of n independent standard normal random variables $Z_i, i = 1, \ldots, n$:

$$X_n = \sum_{i=1}^{n} Z_i^2.$$

The χ^2 statistic

Suppose that when a random experiment is performed, observations may fall into any of n distinct categories. Assuming the truth of a particular hypothesis, H_0, let the probability be p_i that any observation falls in category i. If there are N observations altogether, the expected number, under H_0, that fall in category i is Np_i. We may compare this with the number, N_i, of observations that actually do fall in category i (N_i is random, N is not). To obtain an overall view of how well the observed data fits the model (H_0) we compute the sum of the n squares of the deviations of the N_i from the Np_i, each term in the sum being divided by the **expected number** Np_i. Thus the goodness of fit test statistic is the random variable

$$D_n = \sum_{i=1}^{n} \frac{(N_i - Np_i)^2}{Np_i}.$$

When N is large the random variable D_n has approximately the same probability distribution as X, a χ^2-random variable whose number of degrees of freedom is determined as described below. We therefore put

$$\chi^2 = \sum_{i=1}^{n} \frac{(n_i - Np_i)^2}{Np_i}, \tag{1.10}$$

where n_i is the observed value of N_i, and call (1.10) the value of the χ^2-**statistic**.

If there is close agreement between the observed values (n_i) and those predicted under $H_0(Np_i)$, then the values of $(N_i - Np_i)^2$ and hence D_n will be small. Large observed values of the χ^2-statistic therefore make us inclined to think that H_0 is false.

Critical values of X_n, denoted by $\chi_{n,\alpha}^2$, are defined as follows:

$$\Pr\{X_n > \chi_{n,\alpha}^2\} = \alpha$$

If the value of χ^2 obtained in an experiment is less than the critical value, it is argued that the differences between the values of N_i and Np_i are not large enough to warrant rejecting H_0. On the other hand, if χ^2 exceeds the critical value, H_0 is considered unlikely and is rejected. Often we put $\alpha = .05$, which means that 5% of the time, values of χ^2 greater than the critical value occur even when H_0 is true. That is, there is a 5% chance that we will (incorrectly) reject H_0 when it is true.

In applying the above χ^2 goodness of fit test, the number of degrees of freedom is given by the number n, of 'cells', minus the number of linear relations between the N_i. (There is at least one, $\sum N_i = N$.) The number of degrees of freedom is reduced further by one for each estimated parameter needed to describe the distribution under H_0.

It is recommended that the expected numbers of observations in each category should not be less than 5, but this requirement can often be relaxed. A table of critical values of χ^2 is given in the Appendix, p. 219.

For a detailed account of hypothesis testing and introductory statistics generally, see for example Walpole and Myers (1985), Hogg and Craig (1978) and Mendenhall, Scheaffer and Wackerly (1981). For full accounts of basic probability theory see also Chung (1979) and Feller (1968). Two recent books on applications of probability at an undergraduate level are those of Ross (1985) and Taylor and Karlin (1984).

1.8 NOTATION

Little o

A quantity which depends on Δx but vanishes more quickly than Δx as $\Delta x \to 0$ is said to be 'little o of Δx', written $o(\Delta x)$. Thus for example $(\Delta x)^2$ is $o(\Delta x)$ because $(\Delta x)^2$ vanishes more quickly than Δx. In general, if

$$\lim_{\Delta x \to 0} \frac{g(\Delta x)}{\Delta x} = 0,$$

we write

$$g(\Delta x) = o(\Delta x).$$

The little o notation is very useful to abbreviate expressions in which terms will not contribute after a limiting operation is taken. To illustrate, consider the Taylor expansion of $e^{\Delta x}$:

$$e^{\Delta x} = 1 + \Delta x + \frac{(\Delta x)^2}{2!} + \frac{(\Delta x)^3}{3!} + \cdots$$

$$= 1 + \Delta x + o(\Delta x).$$

We then have

$$\frac{d}{dx} e^x \bigg|_{x=0} = \lim_{\Delta x \to 0} \frac{e^{\Delta x} - 1}{\Delta x}$$

$$= \lim_{\Delta x \to 0} \frac{1 + \Delta x + o(\Delta x) - 1}{\Delta x}$$

$$= \lim_{\Delta x \to 0} \frac{\Delta x}{\Delta x} + \frac{o(\Delta x)}{\Delta x}$$

$$= 1.$$

Equal by definition

As seen already, when we write, for example,

$$q \doteq (1 - p)$$

we are defining the symbol q to be equal to $1 - p$. This is not to be confused with approximately equal to, which is indicated by $\simeq$.

Unit step function

The **unit** (or **Heaviside**) step function located at x_0 is

$$H(x - x_0) = \begin{cases} 0, & x < x_0, \\ 1, & x \geqslant x_0. \end{cases}$$

Thus $H(x - x_0)$ has a jump of $+1$ at x_0 and it is **right-continuous**.

i.i.d.

As seen already, the letters i.i.d. stand for independent and identically distributed.

Probability

Usually the probability of an event A is written

$$\Pr\{A\}$$

but occasionally we just write

$$P\{A\}.$$

REFERENCES

Blake, I.F. (1979). *An Introduction to Applied Probability.* Wiley, New York.

Chung, K.L. (1979). *Elementary Probability Theory.* Springer-Verlag, New York.

Feller, W. (1968). *An Introduction to Probability Theory and its Applications.* Wiley, New York.

Hogg, R.V. and Craig, A.T. (1978). *Introduction to Mathematical Statistics.* Macmillan, New York.

Mendenhall, W., Scheaffer, R.L. and Wackerly, D.D. (1981). *Mathematical Statistics with Applications.* Duxbury, Boston.

Ross, S.M. (1985). *Introduction to Probability Models*. Academic Press, New York.
Taylor, H.M. and Karlin, S. (1984). *An Introduction to Stochastic Modeling*. Academic Press, New York.
Walpole, R.E. and Myers, R.H. (1985). *Probability and Statistics for Engineers and Scientists*. Macmillan, New York.

2

Geometric probability

There are many interesting and challenging problems in applied probability where geometrical considerations are important. We will first look at one which was formulated over 200 years ago by Buffon and has become known as **Buffon's needle problem**. We will then consider two variations on the problem of finding the distribution of the distance between two points whose positions are chosen randomly. In the simpler situation we 'throw' two points onto a line segment. Then we consider the more difficult case of two points which may occur anywhere within a circle. The latter problem has an interesting application in cellular biology.

2.1 BUFFON'S NEEDLE PROBLEM

Buffon's original formulation of this problem was as follows:

I assume that in a room, the floor of which is merely divided by parallel lines, a stick is thrown upwards and one of the players bets the stick will not intersect any of the parallels on the floor, whereas on the contrary the other one bets the stick will intersect some one of these lines; it is required to find the chances of the two players. It is possible to play this game with a sewing needle or headless pin.

Consider Fig. 2.1 where parallel lines are shown with distance D between them. The needle of length L lands as shown.

It is clear that only the ratio of the needle length to the distance between the

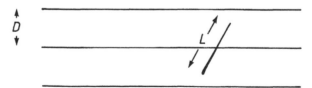

Figure 2.1 The needle is tossed and lands as shown.

lines is relevant, so we let the distance between the lines be unity. (If you are not convinced, work the following derivation through in the general case, and find the claim is true.) Thus L is now the ratio of needle length to the distance between the lines. If $L < 1$, only one intersection is possible and we consider this case.

Theorem 2.1 The probability p that the needle of length $L < 1$ intersects a line is

$$p = \frac{2L}{\pi}.$$

Proof We can specify the needle's orientation by giving the observed values of two random variables. These are:

(i) X, the distance from the lower end, O, of the needle to the nearest line
  above (see Fig. 2.2), and,
(ii) Θ, the angle from the vertical to the needle.

We have to state mathematically what we mean by 'the needle lands randomly'. Our interpretation is that both X and Θ have uniform distributions and that X and Θ are independent.

The possible values of X are between 0 and 1. Thus X is **uniformly distributed** on $(0, 1)$ and has probability density

$$f_X(x) = \begin{cases} 1, & 0 \leqslant x \leqslant 1, \\ 0, & \text{otherwise.} \end{cases}$$

The possible values of Θ are between $-\pi/2$ and $\pi/2$ (O is the lower end) so Θ

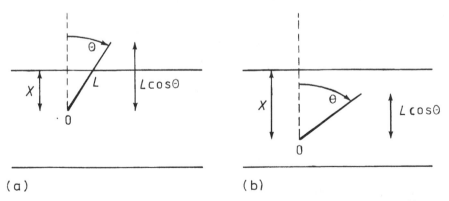

(a)   (b)

Figure 2.2 Random variables X and Θ introduced to solve the needle problem: (a) intersection, (b) no intersection.

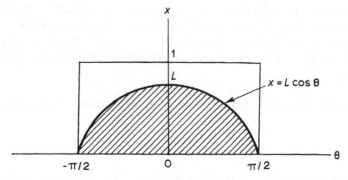

Figure 2.3 The hatched area represents values of X and Θ which lead to an intersection.

is uniformly distributed on $(-\pi/2, \pi/2)$. Its density is therefore

$$f_\Theta(\theta) = \begin{cases} 1/\pi, & -\pi/2 \leqslant \theta \leqslant \pi/2, \\ 0, & \text{otherwise.} \end{cases}$$

Inspection of Fig. 2.2 reveals that

$$p = \Pr\{\text{needle intersects a line}\} = \Pr\{X < L\cos\Theta\}.$$

Since X and Θ are independent, their joint density is given by the product of the densities of X and Θ. Thus,

$$f_{X\Theta}(x, \theta)\,dx\,d\theta = f_X(x)f_\Theta(\theta)\,dx\,d\theta$$

$$= dx\,d\theta/\pi, \qquad 0 \leqslant x \leqslant 1, \quad -\pi/2 \leqslant \theta \leqslant \pi/2.$$

Refer now to Fig. 2.3. The curve consists of points such that $x = L\cos\theta$. Hence we obtain p by integrating the joint density over the hatched area:

$$p = \int_{\theta = -\pi/2}^{\pi/2} \int_{x = 0}^{L\cos\theta} f_{X\Theta}(x, \theta)\,dx\,d\theta$$

$$= \frac{1}{\pi} \int_{\theta = -\pi/2}^{\pi/2} \int_{x = 0}^{L\cos\theta} dx\,d\theta$$

$$= \frac{L}{\pi} \int_{-\pi/2}^{\pi/2} \cos\theta\,d\theta$$

$$= \frac{L}{\pi}[\sin\theta]_{-\pi/2}^{\pi/2} = \frac{2L}{\pi}.$$

This completes the proof.

It is possible to estimate π by performing the following experiment many times.

Throw a 'needle' of length $L < 1$ onto a grid of parallel lines, distance 1 apart. Then the long-run fraction of intersections will approximate p, and π can be estimated by the value of $2L/p$. (See also Theorem 6.13 and subsequent discussion.)

When the needle length is greater than the distance between the lines the possibility arises of more than one intersection (see Exercise 8). For further reading on this and related topics, see Miles and Serra (1978).

2.2 THE DISTANCE BETWEEN TWO RANDOM POINTS ON A LINE SEGMENT

The problem of choosing two points randomly on the unit interval $(0, 1)$ occurs in many different contexts (see for example Exercises 7 and 14). Let the positions of the points be X and Y. The distance between them is then $Z = |X - Y|$. It is assumed that X and Y are independent and uniformly distributed on $(0, 1)$. What is the probability density function of Z?

Theorem 2.2 The density f_Z of Z is

$$f_Z(z) = 2(1 - z), \qquad 0 \leqslant z \leqslant 1.$$

Proof The joint density of X and Y is given by

$$f_{XY}(x, y) = 1, \qquad 0 \leqslant x, y \leqslant 1.$$

Refer to Fig. 2.4. We see that Z will be in the interval $(z, z + dz)$ when (X, Y) falls in the hatched regions, designated A_1 and A_2.

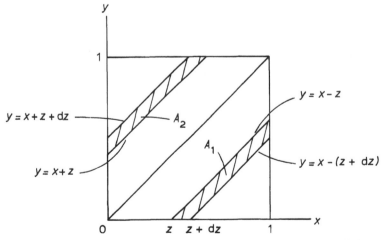

Figure 2.4 The unit square and the regions A_1 and A_2 in which $|X - Y| \in (z, z + dz)$.

We find

$$\Pr\{Z\in(z, z+dz)\} = \iint\limits_{A_1\smile A_2} f_{XY}(x, y)\,dx\,dy$$

$$= 2\int_{x=z}^{1}\left(\int_{y=x-(z+dz)}^{x-z} dy\right)dx,$$

the factor of 2 coming from symmetry considerations. On doing the y-integration,

$$\Pr\{Z\in(z, z+dz)\} = 2\,dz\int_z^1 dx$$

$$= 2\,dz[x]_z^1 = 2(1-z)\,dz.$$

Hence, from (1.3),

$$f_Z(z) = 2(1-z),$$

which gives the required density.

The most likely value of Z is

$$\mathrm{mode}\,(Z) = 0.$$

(You will see this if you sketch the graph of $f_Z(z)$.)

The expected value of Z is

$$E(Z) = \int_0^1 z\,f_Z(z)\,dz$$

$$= 2\int_0^1 z(1-z)\,dz$$

$$= 2\left[\frac{z^2}{2} - \frac{z^3}{3}\right]_0^1 = 2(\tfrac{1}{2} - \tfrac{1}{3}) = 1/3.$$

The second moment of Z is

$$E(Z^2) = \int_0^1 z^2\,f_Z(z)\,dz$$

$$= 2\int_0^1 z^2(1-z)\,dz$$

$$= 2\left[\frac{z^3}{3} - \frac{z^4}{4}\right]_0^1 = 1/6.$$

Hence the **variance** of Z is

$$\mathrm{Var}\,(Z) = \tfrac{1}{6} - \tfrac{1}{9}$$

$$= \tfrac{1}{18}.$$

2.3 THE DISTANCE BETWEEN TWO POINTS DROPPED RANDOMLY IN A CIRCLE

We now turn to a more challenging problem. Two points, $P_1 = (R_1, \Theta_1)$ and $P_2 = (R_2, \Theta_2)$ are dropped randomly in a circle of radius r, as shown in Fig. 2.5. If the distance between them is R, what is the probability density of R?

This problem arises in biology in the following way. In the interior of most cells are chromosomes which carry genetic (hereditary) information (see Fig. 8.1). During the process of mitosis, a cell divides into two copies of itself (see Fig. 10.1). In diploid organisms, of which humans are an example, the chromosomes occur in pairs but usually members of each pair are not distinguishable. However, during the early stages of mitosis, the chromosomes become distinguishable before replication takes place. The question arises whether the positioning of the two members of a pair of chromosomes is random or if there is a force of attraction or repulsion.

Snapshots of a cell undergoing mitosis are called **karyographs** and these may be examined to test the hypothesis of randomness. Under this hypothesis, the distance between the members of a chromosome pair should have approximately the probability distribution which we shall derive.

We assume that the positions of the two points are independent. Randomness is assumed to imply that the probability that P_1 (or P_2) falls in an element of area dA is proportional to dA. Thus

$$\Pr\{R_i \in (r_i, r_i + dr_i), \Theta_i \in (\theta_i, \theta_i + d\theta_i)\} = \frac{r_i\,dr_i\,d\theta_i}{\pi r^2}, \qquad i = 1, 2.$$

We will prove the following result.

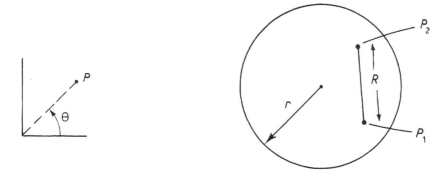

Figure 2.5 Two points P_1 and P_2 are dropped randomly in a circle of radius r, the distance between them being R. The polar coordinates of the two points are (R_1, Θ_1) and (R_2, Θ_2), where angles are measured relative to some reference line.

Theorem 2.3 **The probability density $f_R(\rho)$ of the distance R between two points dropped randomly in a circle of radius r is**

$$f_R(\rho) = \frac{2\rho}{\pi r^2}\left[2\arccos\left(\frac{\rho}{2r}\right) - \frac{\rho}{r}\sqrt{1 - \left(\frac{\rho}{2r}\right)^2}\right], \qquad 0 < \rho < 2r. \quad (2.1)$$

Proof Let the event {The distance between two points dropped randomly in a circle of radius r is in the interval $(\rho, \rho + d\rho)$} be denoted simply by {r}. We will let r vary for fixed ρ and $d\rho$ and hence derive a differential equation for the probability, $P\{r\}$, of this event.

Consider Fig. 2.6 where two circles of radii r and $r + dr$ are shown, the annulus between them being S, the inner circle being marked C and the outer one C'.

A little thought will convince the reader that the following are two mutually exclusive ways in which the two points may be inside C':

(i) both points are inside C;
(ii) at least one point is in S.

Therefore, by the theorem of total probability, the probability that the distance between two points dropped randomly in a circle of radius $r + dr$ is in the interval $(\rho, \rho + d\rho)$ is

$$P\{r + dr\} = P\{r + dr \mid \text{both points are in } C\} \Pr\{\text{both points are in } C\}$$

$$+ P\{r + dr \mid \text{at least one point is in } S\} \Pr\{\text{at least one point is in } S\}.$$
$$(2.2)$$

The four quantities on the right are found as follows. Firstly,

$$P\{r + dr \mid \text{both points are in } C\} = P\{r\}.$$

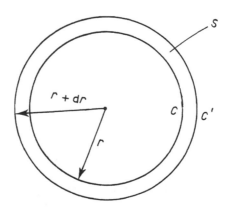

Figure 2.6

Secondly, considering each point separately,

$$\Pr\{\text{either point is in } C\} = \frac{\text{area of } C}{\text{area of } C'}$$

$$= \frac{\pi r^2}{\pi (r + dr)^2}$$

$$= \frac{1}{1 + 2\,dr/r + dr^2/r^2}$$

$$= 1 - \frac{2\,dr}{r} + o(dr).$$

By independence,

$$\Pr\{\text{both points are in } C\} = (1 - \frac{2\,dr}{r} + o(dr))^2$$

$$= 1 - \frac{4\,dr}{r} + o(dr). \qquad (2.3)$$

Thirdly, consider Fig. 2.7. Given at least one point is in S, the probability that the distance separating the points is between ρ and $\rho + d\rho$ is the area of S' divided by the area of C. It is shown in the exercises (Exercise 12) that this is

$$P\{r + dr | \text{at least one point is in } S\} = \frac{2\rho\,d\rho}{\pi r^2} \arccos\left(\frac{\rho}{2r}\right).$$

Finally, because the events 'both points are in C' and 'at least one point is in S' are mutually exclusive and one of these events must occur, we

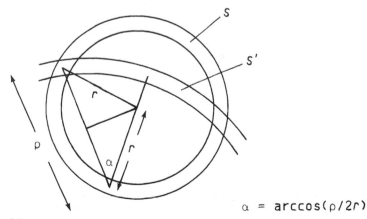

$$\alpha = \arccos(\rho/2r)$$

Figure 2.7

have from (2.3):

$$\Pr\{\text{at least point is in } S\} = \frac{4\,dr}{r} + o(dr).$$

Putting these four calculated quantities in (2.2) gives

$$P\{r + dr\} = P\{r\}\left(1 - \frac{4\,dr}{r}\right) + \frac{2\rho\,d\rho}{\pi r^2}\arccos\left(\frac{\rho}{2r}\right)\frac{4\,dr}{r} + o(dr).$$

Then

$$dP = P\{r + dr\} - P\{r\} = \left[\frac{-4P}{r} + \frac{8\rho\,d\rho}{\pi r^3}\arccos\left(\frac{\rho}{2r}\right)\right]dr + o(dr).$$

Multiplying by r^4 and rearranging gives

$$r^4\,dP + 4r^3P\,dr = \frac{8\rho\,d\rho r}{\pi}\arccos\left(\frac{\rho}{2r}\right)dr + o(dr).$$

'Dividing' by dr we obtain

$$\frac{d}{dr}(Pr^4) = \frac{8\rho\,d\rho r}{\pi}\arccos\left(\frac{\rho}{2r}\right).$$

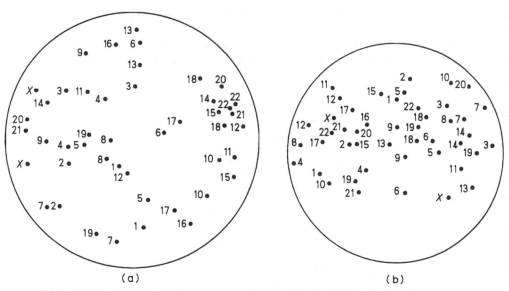

(a)   (b)

Figure 2.8 Karyographs of two normal human female cell nuclei showing chromosome pairs. Note the two X chromosomes. From Barton, David and Fix (1963). Reproduced with permission from *Biometrika* and the authors.

On integrating and rearranging,

$$Pr^4 = \frac{4\rho^2\,d\rho}{\pi} \int \left(\frac{2r}{\rho}\right) \arccos\left(\frac{\rho}{2r}\right) dr + C$$

where C is a constant of integration.

The integration is left as an exercise (Exercise 12). The constant C will be found to be zero by means of the **normalization condition** that the integral of $\int_R(\rho)$ between 0 and $2r$ be unity (recall that $P = \int_R(\rho)\,d\rho$). The theorem follows.

The method of establishing this result has been adopted from Solomon (1978) who attributes the result to Barton, David and Fix (1963). See also Bartlett (1964).

Figure 2.8 shows some karyographs for two normal female cells. Each provides 23 data points for the distance under consideration. Analysis of such pictures in conjunction with distributions obtained under hypotheses of attraction and repulsion indicated that the pairs of chromosomes of the normal female cell were not randomly positioned, whereas those of normal male cells were. Without knowledge of the null distribution (2.3.1), such conclusions could not be accurately drawn. For further details see the references.

2.4 SUM OF TWO RANDOM VARIABLES

The material of this section is not geometric probability but it is not usually covered in introductory probability courses so it did not belong in Chapter 1.

We will need to determine the density of the sum of two random variables in the exercises of Chapter 2 and at other places throughout this book. The following result is obtained for random variables taking values in $(0, \infty)$ and can be modified in other circumstances.

Theorem 2.4   Let X and Y be two independent random variables taking values in $(0, \infty)$ and with densities f_X and f_Y. Let

$$Z = X + Y.$$

Then the density of Z is

$$f_Z(z) = \int_0^z f_X(x) f_1(z - x)\,dx.$$

Proof Consider Fig. 2.9. It can be seen that Z will lie in $(z, z + \Delta z)$ if the pair (X, Y) lies in the region marked S. Thus

$$\Pr\{z < Z < z + \Delta z\} = \iint_S f_{XY}(x, y)\,dx\,dy.$$

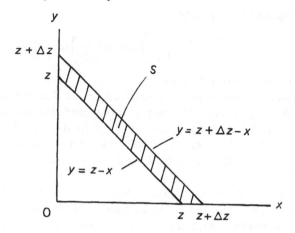

Figure 2.9

Because X and Y are independent, f_{XY}, their joint density, is the product $f_X f_Y$. Considering the region of integration gives

$$\Pr\{z < Z < z + \Delta z\} = \int_{x=0}^{z+\Delta z} f_X(x) \int_{y=z-x}^{z+\Delta z-x} f_Y(y)\,dy\,dx$$

$$= \int_0^{z+\Delta z} f_X(x)\Big[F_Y(z - x + \Delta z) - F_Y(z - x)\Big]dx$$

where F_Y is the distribution function of Y. But if a function g is differentiable at x, we have from elementary calculus that

$$g(x + \Delta x) = g(x) + \Delta x\, g'(x) + o(\Delta x).$$

Thus

$$\Pr\{z < Z < z + \Delta z\} = \int_0^{z+\Delta z} f_X(x)[F_Y(z - x) + \Delta z F_Y'(z - x)$$

$$+ o(\Delta z) - F_Y(z - x)]dx.$$

Now, by definition, the density of Z is

$$f_Z(z) = \lim_{\Delta z \to 0} \frac{\Pr\{z < Z < z + \Delta z\}}{\Delta z}$$

$$= \lim_{\Delta z \to 0} \frac{1}{\Delta z} \int_0^{z+\Delta z} f_X(x)[\Delta z f_Y(z - x) + o(\Delta z)]dx$$

$$= \int_0^z f_X(x) f_Y(z - x)\,dx,$$

as required.

REFERENCES

Bartlett, M.S. (1964). Spectral analysis of two-dimensional point processes. *Biometrika*, **51**, 299-311.

Barton, D.E., David, F.N. and Fix, E. (1963). Random points in a circle and the analysis of chromosome patterns. *Biometrika*, **50**, 23-29.

Blake, I.F. (1979). *An Introduction to Applied Probability*. Wiley, New York.

Miles, R.E. and Serra, J. (eds.) (1978). *Geometrical Probability and Biological Structures: Buffon's 200th Anniversary*. Biomathematics 23. Springer-Verlag, Berlin.

Persson, O. (1964). Distance methods. *Studia Forestalia Svecica*, **15**, 68.

Solomon, H. (1978). *Geometric Probability*. SIAM, Philadelphia.

EXERCISES

1. A point $P = (X, Y)$ is chosen randomly on the unit square. Calculate the probability that P is closer to $(0, 0)$ than it is to $(\frac{1}{2}, \frac{1}{2})$. (*Ans*: 1/8).

2. If X is uniformly distributed on $(0, 1)$ and Y is uniformly distributed on $(0, 2)$, find the probabilities that
   (i) $X + Y > 1$
   (ii) $Y \leqslant X^2 + 1$.
   Assume X and Y are independent. (*Ans*: 3/4; 2/3).

3. Two points P_1 and P_2 are chosen independently and uniformly on $(0, 1)$. What is the probability that the distance between P_1 and P_2 is less than the distance from 0 to P_1? (*Ans*: 3/4).

4. Two numbers, X and Y, are chosen independently and uniformly on $(0, 1)$. Show that the probability that their sum is greater than one while the sum of their squares is less than one is
$$p = \pi/4 - 1/2.$$

5. A carnival game consists of throwing a coin onto a table top marked with contiguous squares of side a. If the coin does not touch any line the player wins. If the coin has radius r, show that the probability of the player's winning is
$$p = (1 - 2r/a)^2.$$

6. Let A and B be jointly uniformly distributed over the quarter-circle $A^2 + B^2 < 1, A > 0, B > 0$. Show that the probability that the equation
$$x^2 + 2A^{1/2}x - (B - 1) = 0$$
has no real roots is
$$p = 2/\pi.$$

7. Two people agree to meet between 12 noon and 1 p.m., but both forget the exact time of their appointment. If they arrive at random and wait for only ten minutes for the other person to show, prove that the probability of

their meeting is

$$p = 11/36.$$

8. Consider Buffon's needle problem with parallel lines distance 1 apart and a needle of length L, $1 < L < 2$. Show that the probability of one intersection is

$$p_1 = \frac{2}{\pi}(L + 2\arccos(1/L) - 2(L^2 - 1)^{1/2}),$$

and the probability of two intersections is

$$p_2 = \frac{2}{\pi}((L^2 - 1)^{1/2} - \arccos(1/L)).$$

9. (Laplace's extension of the Buffon needle problem.) A needle of length L is thrown at random onto a rectangular grid. The rectangles of the grid have sides A and B where $A, B > L$.

(a) Find the expected number of intersections.
(b) Show that the probability that the needle intersects a grid line is

$$p = \frac{2L(A + B) - L^2}{\pi A B}.$$

10. Let X and Y be independent random variables, both exponentially distributed with means $1/\lambda_1$ and $1/\lambda_2$. Find the density of $Z = X + Y$.

11. Let U, V, W be independent random variables taking values in $(0, \infty)$. Show that the density of $Y = U + V + W$ is

$$f_Y(y) = \int_0^y \int_0^z f_U(u) f_V(z - u) f_W(y - z) \, du \, dz.$$

12. With reference to the dropping of two points randomly in a circle,

(i) Show $P\{r + dr | \text{at least one point is in } S\} = \dfrac{2\rho \, d\rho}{\pi r^2} \arccos\left(\dfrac{\rho}{2r}\right).$

(ii) Complete the proof that the density of R is given by (2.1).
(iii) Sketch the graph of f_R given by (2.1).

13. A point is chosen at random within a square of unit side. If U is the square of the distance from the point to the nearest corner of the square, show that the distribution function of U is

$$F_U(u) = \begin{cases} \pi u, & 0 \leqslant u < \tfrac{1}{2}, \\[2mm] 2u \arcsin\left(\dfrac{1 - 2u}{2u}\right) + \sqrt{4u - 1}, & \tfrac{1}{4} \leqslant u < \tfrac{1}{2}, \\[2mm] 1, & u \geqslant \tfrac{1}{2}. \end{cases}$$

(Persson, 1964).

14. Let X_1, X_2, Y_1, Y_2 be independent and uniform on $(0, 1)$. Let $X = |X_2 - X_1|$ and $Y = |Y_2 - Y_1|$. We have seen in Section 2.2 that the densities of X and Y are

$$f_X(x) = 2(1 - x), \qquad f_Y(y) = 2(1 - y), \qquad 0 \leqslant x, y \leqslant 1.$$

(a) Show that the densities of $U = X^2$ and $V = Y^2$ are

$$f_U(u) = u^{-1/2} - 1, \qquad f_V(v) = v^{-1/2} - 1, \qquad 0 \leqslant u, v \leqslant 1.$$

(b) Show that the density of $Z = U + V$ is

$$f_Z(z) = \begin{cases} \pi + z - 4z^{1/2}, & 0 < z < 1 \\ 2\{\arctan(z - 1)^{-1/2} + 2(z - 1)^{1/2} \end{cases}$$

$$- \arctan(z - 1)^{1/2} - 1 - \tfrac{1}{2}z\}, \qquad 1 < z \leqslant \sqrt{2}.$$

(c) Hence show that the density of the distance R between two points dropped randomly in the unit square is

$$f_R(r) = \begin{cases} 2r(\pi + r^2 - 4r), & 0 < r < 1, \\ 4r\{\arctan(r^2 - 1)^{-1/2} + 2(r^2 - 1)^{1/2} \end{cases}$$

$$- \arctan(r^2 - 1)^{1/2} - 1 - r^2/2\}, \qquad 1 \leqslant r \leqslant \sqrt{2}.$$

Note: $\displaystyle \int x^{-1/2}(z - x)^{-1/2} \, dx = 2 \arctan(x/(z - x))^{1/2}.$

Note. Problems 4, 5 and 6 were taken from Blake (1979).

3
Some applications of the hypergeometric and Poisson distributions

In this chapter we will consider some practical situations where the hypergeometric and Poisson distributions arise. We will first consider a technique for estimating animal populations known as **capture–recapture**. This, as we shall see, involves the **hypergeometric distribution**. Poisson random variables arise when we consider randomly distributed points in space or time. One of the applications of this is in the analysis of spatial patterns of plants, which is important in **forestry**. Finally we consider **compound Poisson random variables** with a view to analysing some experimental results in **neurophysiology**.

3.1 THE HYPERGEOMETRIC DISTRIBUTION

The hypergeometric distribution is obtained as follows. A sample of size n is drawn, without replacement, from a population of size N composed of M individuals of type 1 and $N-M$ individuals of type 2. Then the number X of individuals of type 1 in the sample is a **hypergeometric random variable** with probability mass function

$$p_k = \Pr\{X = k\} = \frac{\binom{M}{k}\binom{N-M}{n-k}}{\binom{N}{n}}, \quad \max(0, n - N + M) \leqslant k \leqslant \min(M, n).$$

(3.1)

To derive (3.1) we note that there are $\binom{N}{n}$ ways of choosing the sample of size n from N individuals. The k individuals of type 1 can be chosen from M in $\binom{M}{k}$ ways, and the $n-k$ individuals of type 2 can be chosen from $N-M$ in

$\binom{N-M}{n-k}$ ways. Hence there are $\binom{M}{k}\binom{N-M}{n-k}$ distinct samples with k individuals of type 1 and so (3.1) gives the proportion of samples of size n which contain k individuals of type 1.

The range of X is as indicated in (3.1) as the following arguments show. Recall that there are $N - M$ type 2 individuals. If $n \leqslant N - M$ all members of the sample can be type 2 so it is possible that there are zero type 1 individuals. However, if $n > N - M$, there must be some, and in fact at least $n - (N - M)$, type 1 individuals in the sample. Thus the smallest possible value of X is the larger of 0 and $n - N + M$. Also, there can be no more than n individuals of type 1 if $n \leqslant M$ and no more than M if $M \leqslant n$. Hence the largest possible value of X is the smaller of M and n.

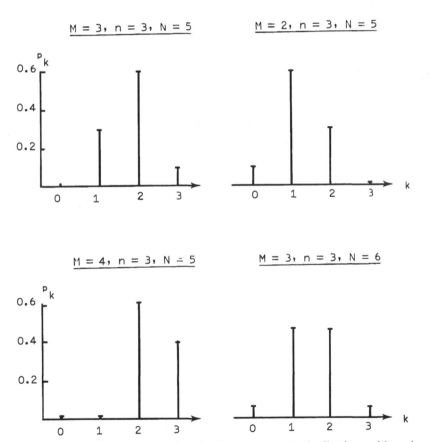

Figure 3.1 Probability mass functions for hypergeometric distributions with various values of the parameters N, M and n.

For the curious we note that the distribution is called hypergeometric because the corresponding generating function is a hypergeometric series (Kendall and Stuart, 1958).

The shape of the hypergeometric distribution depends on the values of the parameters N, M and n. Some examples for small parameter values are shown in Fig. 3.1. Tables are given in Liebermann and Owen (1961).

Mean and variance

The mean of X is

$$E(X) = \frac{nM}{N}$$

and its variance is

$$\operatorname{Var}(X) = \frac{nM(N-n)(N-M)}{N^2(N-1)}. \tag{3.2}$$

Proof We follow the method of Moran (1968). Introduce the **indicator random variables** defined as follows. Let

$$X_i = \begin{cases} 1, & \text{if the } i\text{th member of the sample is type 1,} \\ 0, & \text{otherwise.} \end{cases}$$

Then the total number of type 1 individuals is

$$X = \sum_{i=1}^{n} X_i.$$

Each member of the sample has the same probability of being type 1. Indeed,

$$\Pr\{X_i = 1\} = \frac{M}{N}, \qquad i = 1, \ldots, n. \tag{3.3}$$

as follows from the probability law (3.1) when $k = n = 1$.
  Since

$$E(X_i) = \frac{M}{N},$$

we see that

$$E(X) = nE(X_i) = \frac{nM}{N}.$$

To find $\text{Var}(X)$ we note that the second moment of X is

$$E(X^2) = E\left(\left(\sum_{i=1}^{n} X_i\right)^2\right) = E\left(\sum_{i=1}^{n}\sum_{j=1}^{n} X_i X_j\right)$$

$$= E\left(\sum_{i=1}^{n} X_i^2 + \sum_{\substack{i=1 \\ i \neq j}}^{n}\sum_{j=1}^{n} X_i X_j\right). \qquad (3.4)$$

The expected value of X_i^2 is, from (3.3),

$$E(X_i^2) = \frac{M}{N}, \qquad i = 1, 2, \ldots, n. \qquad (3.5)$$

We now use (3.1) with $n = k = 2$, to get the probability that $X = 2$ when $n = 2$. This gives

$$\Pr\{X_i = 1, X_j = 1\} = \frac{\binom{M}{2}\binom{N-M}{0}}{\binom{N}{2}}$$

$$= \frac{M(M-1)}{N(N-1)}, \qquad i,j = 1, 2, \ldots, n, \qquad i \neq j.$$

Thus

$$E(X_i X_j) = \frac{M(M-1)}{N(N-1)}, \qquad (3.6)$$

and there are $n(n-1)$ such terms in (3.4). Substituting (3.5) and (3.6) in (3.4) gives

$$E(X^2) = \frac{nM}{N} + \frac{n(n-1)M(M-1)}{N(N-1)}.$$

Formula (3.2) follows from the relation

$$\text{Var}(X) = E(X^2) - E^2(X).$$

3.2 ESTIMATING A POPULATION FROM CAPTURE-RECAPTURE DATA

Assume now that a population size N is unknown. The population may be the kangaroos or emus in a certain area or perhaps the fish in a lake or stream. We wish to estimate N without counting every individual. A method of doing this is called capture-recapture. Here M individuals are captured, marked in some distinguishing way and then released back into the population. Later, after a satisfactory mixing of the marked and unmarked individuals is attained, a sample of size n is taken from the population and the number X of marked individuals is noted. This method, introduced by Laplace in 1786 to estimate France's population, is often employed by biologists and individuals in resource management to estimate animal populations. The method we have described is called **direct sampling**. Another method (**inverse sampling**) is considered in Exercises 5 and 6.

In the capture–recapture model the marked individuals correspond to type 1 and the unmarked to type 2. The probability that the number of marked individuals is k is thus given by (3.1). Suppose now that k is the number of marked individuals in the sample; then values of N for which $\Pr\{X = k\}$ is very small are considered unlikely. One takes as an estimate, $\hat{N}$, of N that value which maximizes $\Pr\{X = k\}$. $\hat{N}$ is a random variable and is called the **maximum likelihood estimate** of N.

Theorem 3.1 The maximum likelihood estimate of N is

$$\hat{N} = \left[\frac{Mn}{X}\right],$$

where $[z]$ denotes the greatest integer less than z.

Proof We follow Feller (1968). Holding M and n constant we let $\Pr\{X = k\}$ for a fixed value of N be denoted $p_N(k)$. Then

$$\frac{p_N(k)}{p_{N-1}(k)} = \frac{\binom{M}{k}\binom{N-M}{n-k}}{\binom{N}{n}} \Bigg/ \frac{\binom{M}{k}\binom{N-1-M}{n-k}}{\binom{N-1}{n}},$$

which simplifies to

$$\frac{p_N(k)}{p_{N-1}(k)} = \frac{N^2 - MN - nN + Mn}{N^2 - MN - nN + kN}.$$

We see that p_N is greater than, equal to, or less than p_{N-1} according as Mn is greater than, equal to, or less than kN; or equivalently as N is less than, equal to or greater than Mn/k. Excluding for now the case where Mn/k is an integer, the sequence $\{p_N, N = 1, 2, \ldots\}$ is increasing as long as $N < Mn/k$ and is decreasing when $N > Mn/k$. Thus the maximum value of p_N occurs when

$$N = \left[\frac{Mn}{k}\right],$$

which is the largest integer less than Mn/k.

In the event that Mn/k is an integer the maximum value of p_N will be $p_{Mn/k}$ and $p_{Mn/k-1}$, these being equal. One may then use

$$\frac{Mn}{k} - 1 = \left[\frac{Mn}{k}\right]$$

as an estimate of the population. This completes the proof.

Approximate confidence intervals for $\hat{N}$

In situations of practical interest N will be much larger than both M and n. Let us assume in fact that N is large enough to regard the sampling as approximately with replacement. If $\tilde{X}_i$ approximates X_i in this scheme, then, for all i from 1 to n,

$$\Pr\{\tilde{X}_i = 1\} = \frac{M}{N} = 1 - \Pr\{\tilde{X}_i = 0\}.$$

The approximation to X is then given by

$$\tilde{X} = \sum_{i=1}^{n} \tilde{X}_i.$$

This is a binomial random variable with parameters n and M/N so that

$$\Pr\{\tilde{X} = k\} = b\left(k; n, \frac{M}{N}\right), \qquad k = 0, 1, \ldots, n.$$

The expectation and variance of $\tilde{X}$ are

$$E(\tilde{X}) = \frac{nM}{N} = E(X)$$

$$\mathrm{Var}(\tilde{X}) = \frac{nM}{N}\left(1 - \frac{M}{N}\right) \simeq \mathrm{Var}(X).$$

Furthermore, if the sample size n is fairly large, the distribution of $\tilde{X}$ can be approximated by that of a normal random variable with the same mean and variance (see Chapter 6). Replacing N by the observed value, $\hat{n}$, of its maximum likelihood estimator gives

$$\tilde{X} \stackrel{d}{\simeq} N\left(\frac{nM}{\hat{n}}, \sqrt{\frac{nM}{\hat{n}}\left(1 - \frac{M}{\hat{n}}\right)}\right),$$

where $\stackrel{d}{\simeq}$ means 'has approximately the same distribution'.

Ignoring the technicality of integer values, we have

$$\hat{n} = \frac{nM}{k},$$

where k is the observed value of X, so

$$\tilde{X} \stackrel{d}{\simeq} N\left(k, \sqrt{k\left(1 - \frac{k}{n}\right)}\right).$$

Using the standard symbol Z for an $N(0, 1)$ random variable and the usual notation

$$\Pr\{Z > z_{\alpha/2}\} = \frac{\alpha}{2},$$

we find

$$\Pr\left\{k - z_{\alpha/2}\sqrt{k\left(1 - \frac{k}{n}\right)} < \tilde{X} < k + z_{\alpha/2}\sqrt{k\left(1 - \frac{k}{n}\right)}\right\} \simeq 1 - \alpha.$$

However, $\hat{N} = Mn/X$, so we obtain the following approximate result when the observed number of marked individuals in the recaptured sample is k.

Theorem 3.2 An approximate $100(1 - \alpha)\%$ confidence interval for the estimator $\hat{N}$ of the population is

$$\Pr\left\{\frac{nM}{k + z_{\alpha/2}\sqrt{k\left(1 - \frac{k}{n}\right)}} < \hat{N} < \frac{nM}{k - z_{\alpha/2}\sqrt{k\left(1 - \frac{k}{n}\right)}}\right\} \simeq 1 - \alpha.$$

Thus for example, if a 95% confidence interval is required we put $z_{\alpha/2} = z_{.025} = 1.96$ in this formula.

Discussion

The above estimates have been obtained for direct sampling in the ideal situation. Before applying them in any real situation an examination of the assumptions made would be worth while. Among these are:

(i) The marked individuals disperse randomly and homogeneously throughout the population.
(ii) All marked individuals retain their marks.
(iii) Each individual, whether marked or not, has the same chance of being in the recaptured sample.
(iv) There are no losses due to death or emigration and no gains due to birth or immigration.

Some of these assumptions can be relaxed in a relatively simple way (see Exercise 7). In the approach mentioned earlier called inverse sampling, the recapturing takes place until a predetermined number of marked individuals is obtained. For useful refinements of the basic method presented above see

Manly (1984) and references therein; see also Cormack (1968) and the conference article of the same author (1973) who begins with the following remarks:

Many of the papers in this volume are concerned with the process of describing the development of an animal population by a mathematical model. The properties of such a model can then be derived, either by elegant mathematics or equally elegant computer simulation, in order to describe the future state of the population in terms of certain initial boundary conditions. The model becomes of scientific value when such predictions can be tested, which requires in turn that the mathematical symbols can be replaced by numbers. The parameters of the model must be estimated from data of a type that a biologist can collect about the population he is studying.

For an introductory treatment written for biologists, see Begon (1979).

3.3 THE POISSON DISTRIBUTION

We recall the definition and some elementary properties of Poisson random variables.

Definition A non-negative integer-valued random variable X has a Poisson distribution with parameter $\lambda > 0$ if

$$\boxed{p_k = \Pr\{X = k\} = \frac{e^{-\lambda}\lambda^k}{k!}}, \qquad k = 0, 1, 2, \ldots \tag{3.7}$$

From the definition of e^λ as $\sum_0^x \lambda^k/k!$ we find

$$\sum_{k=0}^{\infty} \Pr\{X = k\} = 1.$$

The mean and variance of X will easily be found to be

$$E(X) = \mathrm{Var}(X) = \lambda.$$

The shape of the probability mass function depends on λ as Table 3.1 and the graphs of Fig. 3.2 illustrate.

Table 3.1 Probability mass functions for some Poisson random variables

	p_0	p_1	p_2	p_3	p_4	p_5	p_6	p_7	p_8
$\lambda = \frac{1}{2}$	.607	.303	.076	.013	.002	< .001			
$\lambda = 1$	.368	.368	.184	.061	.015	.003	< .001		
$\lambda = 2$	.135	.271	.271	.180	.090	.036	.012	.003	< .001

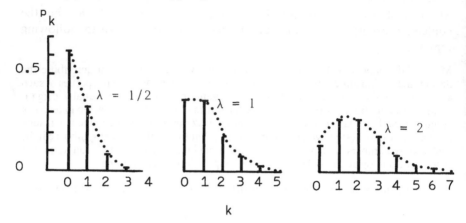

Figure 3.2 Probability mass functions for Poisson random variables with various parameter values.

There are two points which emerge just from looking at Fig. 3.2:

(i) Poisson random variables with different parameters can have quite different looking mass functions.
(ii) When λ gets large the mass function has the shape of a normal density (see Chapter 6).

Poisson random variables arise frequently in counting numbers of events. We will consider events which occur randomly in one-dimensional space or time and in two-dimensional space, the latter being of particular relevance in ecology. Generalizations to higher-dimensional spaces will also be briefly discussed.

3.4 HOMOGENEOUS POISSON POINT PROCESS IN ONE DIMENSION

Let t represent a time variable. Suppose an experiment begins at $t = 0$. Events of a particular kind occur randomly, the first being at T_1, the second at T_2, etc., where T_1, T_2, etc., are random variables. The values t_i of $T_i, i = 1, 2, \ldots$ will be called **points of occurrence** or just *events* (see Fig. 3.3).

Figure 3.3 Events of a particular kind occur at t_1, t_2, etc.

Let $(s_1, s_2]$ be a subinterval of the interval $[0, s]$ where $s < \infty$. Denote by $N(s_1, s_2)$ the number of points of occurrence in $(s_1, s_2]$. Then $N(s_1, s_2)$ is a random variable and the collection of all such random variables, abbreviated to N, for various subintervals (or actually any subsets of $[0, s]$) is called a **point process** on $[0, s]$.

Definition N **is an homogeneous Poisson point process with rate λ if:**

(i) **for any $0 \leqslant s_1 < s_2 \leqslant s$, $N(s_1, s_2)$ is a Poisson random variable with parameter $\lambda (s_2 - s_1)$;**
(ii) **for any collection of times $0 \leqslant s_0 < s_1 < s_2 \ldots < s_n \leqslant s$, where $n \geqslant 2$, the random variables $\{N(s_{k-1}, s_k), k = 1, 2, \ldots, n\}$ are mutually independent.**

We see therefore that the number of points of occurrence in $(0, t]$, which we denote by just $N(t)$, is a Poisson random variable with parameter λt. Also, the numbers of points falling in disjoint intervals are independent.

Now, the expected value of $N(t)$ is λt and this is also the expected number of points in $(0, t]$. Thus the expected number of points in the unit interval $(0, 1]$, or any other interval of unit length, is just λ. Hence the description of λ as the rate parameter, or as it is often called, the **intensity** of the process. The process is called homogeneous because the probability law of the number of points in any interval depends only on the length of the interval, not on its location.

We will now derive some properties of interest in connection with the distances (or time intervals) between points of occurrence (events) when the Poisson point process is defined on subsets of $[0, \infty)$. The role of s will now change.

The waiting time to the next event

Consider any fixed time point $s > 0$. Let T_1 be the time which elapses before the first event after s. Then we have the following result.

Theorem 3.3   The waiting time, T_1, for an event is exponentially distributed with mean $1/\lambda$.

Note that the distribution of T_1 does not depend on s. We say the process has no memory, a fact which is traced to the definition since the numbers of events in $(s_1, s_2]$ and $(s_2, s_3]$ are independent.

Proof First we note that the probability of one event in any small interval of length Δt is

$$e^{-\lambda \Delta t}(\lambda \Delta t) = \lambda \Delta t + o(\Delta t), \qquad (3.8)$$

where $o(\Delta t)$ here stands for terms which vanish faster than Δt as Δt goes to zero. We will have $T_1 \in (t, t + \Delta t]$ if there are no events in $(s, s + t]$ and one event in $(s + t, s + t + \Delta t]$. By independence, the probability of both of these is the product of the probabilities of either occurring separately. Hence

$$\Pr\{T_1 \in (t, t + \Delta t]\} = e^{-\lambda t}[\lambda \Delta t + o(\Delta t)].$$

It follows that the density of T_1 is given by

$$\boxed{f_{T_1}(t) = \lambda e^{-\lambda t}, \quad t > 0}$$

Alternatively, this result may be obtained by noting that

$$\Pr\{T_1 > t\} = \Pr\{N(s, s + t) = 0\} = e^{-\lambda t}.$$

We find that not only is the waiting time to an event exponentially distributed but also the following is true.

Theorem 3.4 The time interval between events is exponentially distributed with mean $1/\lambda$.

Proof This is Exercise 8.

In fact it can be shown that if the distances between consecutive points of occurrence are independent and identically exponentially distributed, then the point process is a homogeneous Poisson point process. This statement provides one basis for statistical tests for a Poisson process (Cox and Lewis, 1966).

The waiting time to the kth point of occurrence

Theorem 3.5 Let T_k be the waiting time until the kth event after s, $k = 1, 2, \ldots$. Then T_k has a gamma density with parameters k and λ.

Proof The kth point of occurrence will be the only one in $(s + t, s + t + \Delta t]$ if and only if there are $k - 1$ points in $(s, s + t]$ and one point is in $(s + t, s + t + \Delta t]$. It follows from (3.7) and (3.8) that

$$\Pr\{T_k \in (t, t + \Delta t]\} = \frac{e^{-\lambda t}(\lambda t)^{k-1}[\lambda \Delta t + o(\Delta t)]}{(k - 1)!}, \qquad k = 1, 2, \ldots$$

Hence the density of T_k is

$$\boxed{f_{T_k}(t) = \frac{\lambda (\lambda t)^{k-1} e^{-\lambda t}}{(k - 1)!}, \quad t > 0} \tag{3.9}$$

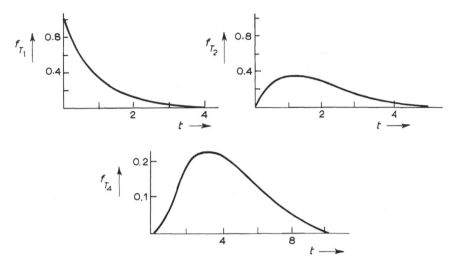

Figure 3.4 The densities of the waiting times for 1, 2 and 4 events in a homogeneous Poisson point process with $\lambda = 1$.

and the mean and variance of T_k are given by

$$E(T_k) = \frac{k}{\lambda}, \quad \text{Var}(T_k) = \frac{k}{\lambda^2}.$$

Note that this result can also be deduced from the fact that the sum of $k > 1$ independent exponentially distributed random variables, each with mean $1/\lambda$, has a gamma density as in (3.9) (prove this by using Theorem 2.4). Furthermore, it can be shown that the waiting time to the kth event after an event has a density given by (3.9).

The waiting times for $k = 1, 2$ and 4 events have densities as depicted in Fig. 3.4. Note that as k gets larger, the density approaches that of a normal random variable (see Chapter 6, when we discuss the central limit theorem).

3.5 OCCURRENCE OF POISSON PROCESSES IN NATURE

The following reasoning leads in a natural way to the Poisson point process. Points, representing the times of occurrence of an event, are sprinkled randomly on the interval $[0, s]$ under the assumptions:

(i) the numbers of points in disjoint subintervals are independent;
(ii) the probability of finding a point in a very small subinterval is

proportional to its length, whereas the probability of finding more than one point is negligible.

It is convenient to divide $[0, s]$ into n subintervals of equal length $\Delta s = s/n$. Under the above assumptions the probability p that a given subinterval contains a point is $\lambda s/n$ where λ is a positive constant. Hence the chance of k occupied subintervals is

$$\Pr\{k \text{ points in } [0, s]\} = b(k; n, p)$$

$$= b\left(k; n, \frac{\lambda s}{n}\right).$$

Now as $n \to \infty$, $\lambda s/n \to 0$ and we may invoke the Poisson approximation to the binomial probabilities (see also Chapter 6):

$$b(k; n, p) \xrightarrow[n \to \infty]{} \frac{\exp(-np)(np)^k}{k!}.$$

But $np = n(\lambda s)/n = \lambda s$. Hence in the limit as $n \to \infty$,

$$\Pr\{k \text{ points in } [0, s]\} = \frac{\exp(-\lambda s)(\lambda s)^k}{k!},$$

as required.

The above assumptions and limiting argument should help to make it understandable why approximations to Poisson point processes arise in the study of a broad range of natural random phenomena. The following examples provide evidence for this claim.

Examples

(i) Radioactive decay
The times at which a collection of atomic nuclei emit, for example, alpha-particles can be well approximated as a Poisson point process. Suppose there are N observation periods of duration T, say. In Exercise 18 it is shown that under the Poisson hypothesis, the expected value, n_k, of the number, N_k, of observation periods containing k emissions is

$$n_k = \frac{N \exp(-\bar{n})\bar{n}^k}{k!}, \qquad k = 0, 1, 2, \ldots \tag{3.10}$$

where $\bar{n} = \lambda T$ is the expected number of emissions per observation period. For an experimental data set, see Feller (1968, p. 160).

(ii) Arrival times
The times of arrival of customers at stores, banks, etc., can often be approximated by Poisson point processes. Similarly for the times at which

phone calls are made, appliances are switched on, accidents in factories or in traffic occur, etc. In queueing theory the Poisson assumption is usually made (see for example Blake, 1979), partly because of empirical evidence and partly because it leads to mathematical simplifications. In most of these situations the rate may vary so that $\lambda = \lambda(t)$. However, over short enough time periods, the assumption that λ is constant will often be valid.

(iii) Mutations

In cells changes in genetic (hereditary) material occur which are called mutations. These may be spontaneous or induced by external agents. If mutations occur in the reproductive cells (gametes) then the offspring inherits the mutant genes. In humans the rate at which spontaneous mutations occur per gene is about 4 per hundred thousand gametes (Strickberger, 1968). In the common bacterium *E. coli*, a mutant variety is resistant to the drug streptomycin. In one experiment, $N = 150$ petri dishes were plated with one million bacteria each. It was found that 98 petri dishes had no resistant colonies, 40 had one, 8 had two, 3 had three and 1 had four. The average number $\bar{n}$ of mutants per million cells (bacteria) is therefore

$$\bar{n} = \frac{40 \times 1 + 8 \times 2 + 3 \times 3 + 1 \times 4}{150} = 0.46.$$

Under the Poisson hypothesis, the expected numbers n_k of dishes containing k mutants are as given in Table 3.2, as calculated using (3.10). The observed values N_k are also given and the agreement is reasonable. This can be demonstrated with a χ^2 test (see Chapter 1).

(iv) Voltage changes at nerve-muscle junction

The small voltage changes seen in a muscle cell attributable to spontaneous activity in neighbouring nerve cells occur at times which are well described as a Poisson point process. A further aspect of this will be elaborated on in Section 3.9. Figure 3.5 shows an experimental histogram of waiting times

Table 3.2 Bacterial mutation data*

k	n_k	N_k(Obs.)
0	94.7	98
1	43.5	40
2	10.0	8
3	1.5	3
4	0.2	1

*From Strickberger (1968).

Figure 3.5 A histogram of waiting times between spontaneously occurring small voltage changes in a muscle cell due to activity in a neighbouring nerve cell. From Fatt and Katz (1952).

between such events. According to the Poisson assumption, the waiting time should have an exponential density which is seen to be a good approximation to the observed data. This may also be rendered more precise with a χ^2 goodness of fit test. For further details see Van der Kloot *et al.* (1975).

3.6 POISSON POINT PROCESSES IN TWO DIMENSIONS

Instead of considering random points on the line we may consider random points in the plane $\mathbb{R}^2 = \{(x, y)| - \infty < x < \infty, - \infty < y < \infty\}$, or subsets thereof.

Definition A point process N is an homogeneous Poisson point process in the plane with intensity λ if:

(i) **for any subset A of $\mathbb{R}^2$, the number of points $N(A)$ occurring in A is a Poisson random variable with parameter $\lambda|A|$, where $|A|$ is the area of A;**

(ii) **for any collection of disjoint subsets of $\mathbb{R}^2$, $A_1, A_2, \ldots, A_n$, the random variables $\{N(A_k), k = 1, 2, \ldots, n\}$ are mutually independent.**

Note that the number of points in $[0, x] \times [0, y]$ is a Poisson random variable with parameter λxy. Putting $x = y = 1$ we find that the number of points in the unit square is Poisson with parameter λ. Hence λ is the expected number of points per unit area.

Application to ecological patterns

Ecologists are interested in the spatial distributions of plants and animals (see for example MacArthur and Connell, 1966). Three of the situations of interest are:

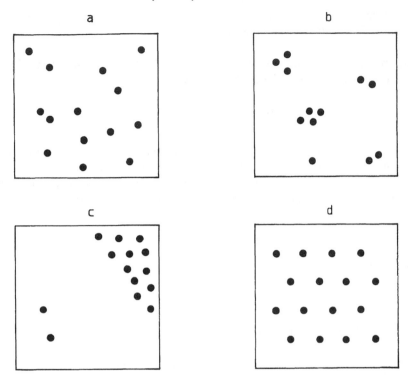

Figure 3.6 Some representative spatial patterns of organisms: (a) random, (b) clumping in groups, (c) preferred location, (d) regular.

(i) the organisms are distributed randomly;
(ii) the organisms have preferred locations in the sense that they tend to occur in groups (i.e. are clustered or clumped) or in some regions more frequently than others;
(iii) the organisms are distributed in a regular fashion in the sense that the distances between them and their nearest neighbours tend to be constant.

These situations are illustrated in Fig. 3.6. We note that clumping indicates cooperation between organisms. The kind of spacing shown in Fig. 3.6(d) indicates competition as the organisms tend to maintain a certain distance between themselves and their neighbours.

An important reason for analysing the underlying pattern is that if it is known, the total population may be estimated from a study of the numbers in a small region. This is of particular importance in the forest industry.

The hypothesis of randomness leads naturally, by the same kind of argument as in Section 3.5, to a Poisson point process in the plane. Ecologists refer to this as a **Poisson forest**. Under the assumption of a Poisson forest we may derive the probability density function of the distance from one organism (e.g. tree) to its nearest neighbour. We may use this density to test the hypothesis of randomness. We first note the following result.

Theorem 3.6 In a Poisson forest, the distance R_1 from an arbitrary fixed point to the nearest event has the probability density

$$\boxed{f_{R_1}(r) = 2\lambda\pi r e^{-\lambda\pi r^2}}, \qquad r > 0. \tag{3.11}$$

Proof We will have $R_1 > r$ if and only if there are no events in the circle of radius r with centre at the fixed point under consideration. Such a circle has area πr^2, so from the definition of a Poisson point process in the plane, the number of events inside the circle is a Poisson random variable with mean $\lambda\pi r^2$. This gives

$$\Pr\{R_1 > r\} = e^{-\lambda\pi r^2}.$$

We must then have

$$f_{R_1}(r) = \frac{\mathrm{d}}{\mathrm{d}r}(1 - e^{-\lambda\pi r^2})$$

which leads to (3.11) as required.

We may also prove that the distance from an event to its nearest neighbour in a Poisson forest has the density given by (3.11). It is left as an exercise to prove the following result.

Theorem 3.7 In a Poisson forest the distance R_k to the kth nearest event has the density

$$\boxed{f_{R_k}(r) = \frac{2\pi\lambda r(\lambda\pi r^2)^{k-1} e^{-\lambda\pi r^2}}{(k-1)!}}, \qquad r > 0, \qquad k = 1, 2, \dots.$$

Estimating the number of trees in a forest

If one is going to estimate the number of trees in a forest, it must first be ensured that the assumed probability model is valid. The obvious hypothesis to begin with is that one is dealing with a Poisson process in the plane. A few methods of testing this hypothesis and a method of estimating λ are now outlined. For some further references see Patil *et al.* (1971) and Heltshe and Ritchey (1984). An actual data set is shown in Fig. 3.7.

Method 1 – Distance measurements
Under the assumption of a Poisson forest the point–nearest tree or tree–nearest tree distance has the density f_{R_1} given in (3.11). The actual measure-

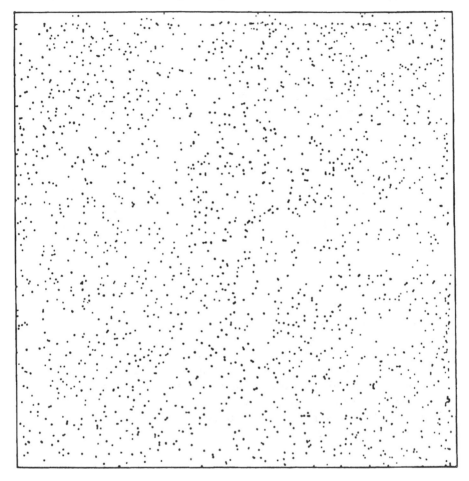

Figure 3.7 Locations of trees in Lansing Woods. Smaller dots represent oaks, larger dots represent hickories and maples. The data are analysed in Exercise 22. Reproduced with permission from Clayton (1984).

ments of such distances may be collected into a histogram or empirical distribution function. A goodness of fit test such as χ^2 (see Chapter 1) or Kolmogorov–Smirnov (see for example Hoel, 1971; or Afifi and Azen, 1979) can be carried out. Note that **edge effects** must be minimized since the density of R_1 was obtained on the basis of an infinite forest.

Assuming a Poisson forest the parameter λ may be estimated as follows. Let $\{X_i, i = 1, 2, \ldots, n\}$ be a random sample for the random variable with the density (3.11). Then it is shown in Exercise 21 that an **unbiased estimator** (see

Exercise 6) of $1/\lambda$ is

$$\hat{\Lambda}^{-1} = \frac{\pi}{n} \sum_{i=1}^{n} X_i^2.$$

An estimate of λ is thus made and hence, if the total area A is known, the total number of trees may be estimated as λA. For further details see Diggle (1975, 1983), Ripley (1981) and Upton and Fingleton (1985).

Method 2 – Counting

Another method of testing the hypothesis of a Poisson forest is to subdivide the area of interest into N equal smaller areas called cells. The numbers N_k of cells containing k plants can be compared using a χ^2-test with the expected numbers under the Poisson assumption using (3.10), with $\bar{n} =$ the mean number of plants per cell.

Extensions to three and four dimensions

Suppose objects are randomly distributed throughout a 3-dimensional region. The above concepts may be extended by defining a Poisson point process in $\mathbb{R}^3$. Here, if A is a subset of $\mathbb{R}^3$, the number of objects in A is a Poisson random variable with parameter $\lambda|A|$, where λ is the mean number of objects per unit volume and $|A|$ is the volume of A. Such a point process will be useful in describing distributions of organisms in the ocean or the earth's atmosphere, distributions of certain rocks in the earth's crust and of objects in space. Similarly, a Poisson point process may be defined on subsets of $\mathbb{R}^4$ with a view to describing random events in space–time.

3.7 COMPOUND POISSON RANDOM VARIABLES

Let $X_k, k = 1, 2, \dots$ be independent identically distributed random variables and let N be a non-negative integer-valued random variable, independent of the X_k. Then we may form the following sum:

$$S_N = X_1 + X_2 + \cdots + X_N, \tag{3.12}$$

where the number of terms is determined by the value of N. Thus S_N is a **random sum of random variables**: we take S_N to be zero if $N = 0$. If N is a Poisson random variable, S_N is called a **compound Poisson random variable**. The mean and variance of S_N are then as follows.

Theorem 3.8 Let $E(X_1) = \mu$ and $\mathrm{Var}(X_1) = \sigma^2$, $|\mu| < \infty, \sigma < \infty$. If N is

Poisson with parameter λ, then S_N defined by (3.12) has mean and variance

$$E(S_N) = \lambda\mu$$
$$\text{Var}(S_N) = \lambda(\mu^2 + \sigma^2).$$

Proof The law of total probability applied to expectations (see p. 8) gives

$$E(S_N) = \sum_{k=0}^{\infty} E(S_N | N = k) \Pr\{N = k\}.$$

But conditioned on $N = k$, there are k terms in (3.12) so $E(S_N | N = k) = k\mu$. Thus

$$E(S_N) = \sum_{k=0}^{\infty} \mu k \Pr\{N = k\}$$

$$= \mu E(N)$$

$$= \lambda\mu.$$

Similarly,

$$E(S_N^2) = \sum_{k=0}^{\infty} E(S_N^2 | N = k) \Pr\{N = k\}$$

$$= \sum_{k=0}^{\infty} [\text{Var}(S_N | N = k) + E^2(S_N | N = k)] \Pr\{N = k\}$$

$$= \sum_{k=0}^{\infty} (k\sigma^2 + k^2\mu^2) \Pr\{N = k\}$$

$$= \sigma^2 E(N) + \mu^2 E(N^2)$$

$$= \sigma^2\lambda + \mu^2[\text{Var}(N) + E^2(N)]$$

$$= \sigma^2\lambda + \mu^2(\lambda + \lambda^2).$$

The result follows since $\text{Var}(S_N) = E(S_N^2) - \lambda^2\mu^2$.

Example The number of seeds (N) produced by a certain kind of plant has a Poisson distribution with parameter λ. Each seed, independently of how many there are, has probability p of forming into a developed plant. Find the mean and variance of the number of developed plants (ignoring the parent).

Solution Let $X_k = 1$ if the kth seed develops into a plant and let $X_k = 0$ if it doesn't. Then the X_k are i.i.d. Bernoulli random variables with

$$\Pr\{X_1 = 1\} = p = 1 - \Pr\{X_1 = 0\}$$

and

$$E(X_1) = p$$
$$\text{Var}(X_1) = p(1 - p).$$

The number of developed plants is

$$S_N = X_1 + X_2 + \cdots + X_N$$

which is therefore a compound Poisson random variable. By Theorem 3.8, with $\mu = p$ and $\sigma^2 = p(1 - p)$ we find

$$E(S_N) = \lambda p$$
$$\mathrm{Var}\,(S_N) = \lambda(p^2 + p(1 - p))$$
$$= \lambda p.$$

As might be suspected from these results, in this example S_N is itself a Poisson random variable with parameter λp. This can be readily shown using generating functions – see Section 10.4.

3.8 THE DELTA FUNCTION

We will consider an interesting neurophysiological application of compound Poisson random variables in the next section. Before doing so we find it convenient to introduce the delta function. This was first employed in quantum mechanics by the celebrated theoretical physicist P.A.M. Dirac, but has since found application in many areas.

Let X_ε be a random variable which is uniformly distributed on $(x_0 - \varepsilon/2, x_0 + \varepsilon/2)$. Then its distribution function is

$$F_{X_\varepsilon}(x) = \mathrm{Pr}\,\{X_\varepsilon \leqslant x\} = \begin{cases} 0, & x \leqslant x_0 - \varepsilon/2, \\ \dfrac{1}{\varepsilon}[x - (x_0 - \varepsilon/2)], & |x - x_0| < \varepsilon/2, \\ 1, & x \geqslant x_0 + \varepsilon/2 \end{cases}$$

$$\doteq H_\varepsilon(x - x_0).$$

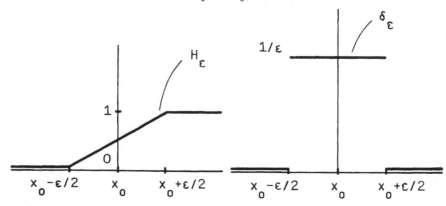

Figure 3.8

The density of X_ε is

$$f_{X_\varepsilon}(x) = \frac{dF_{X_\varepsilon}}{dx} = \begin{cases} 1/\varepsilon, & |x - x_0| < \varepsilon/2, \\ 0, & \text{otherwise.} \end{cases}$$

$$\doteq \delta_\varepsilon(x - x_0).$$

The functions H_ε and δ_ε are sketched in Fig. 3.8.

As $\varepsilon \to 0$, $H_\varepsilon(x - x_0)$ approaches the unit step function, $H(x - x_0)$ and $\delta_\varepsilon(x - x_0)$ approaches what is called a delta function, $\delta(x - x_0)$. In the limit as $\varepsilon \to 0$, δ_ε becomes 'infinitely large on an infinitesimally small interval' and zero everywhere else. We always have for all $\varepsilon > 0$,

$$\int_{-\infty}^{\infty} \delta_\varepsilon(x - x_0)\,dx = 1.$$

We say that the limiting object $\delta(x - x_0)$ is a **delta function** or a **unit mass** concentrated at x_0.

Substitution property

Let f be an arbitrary function which is continuous on $(x_0 - \varepsilon/2, x_0 + \varepsilon/2)$. Consider the integrals

$$I_\varepsilon = \int_{-\infty}^{\infty} f(x)\delta_\varepsilon(x - x_0)\,dx = \frac{1}{\varepsilon} \int_{x_0 - \varepsilon/2}^{x_0 + \varepsilon/2} f(x)\,dx.$$

When ε is very small,

$$I_\varepsilon \simeq \frac{1}{\varepsilon}\varepsilon f(x_0) = f(x_0).$$

We thus obtain the **substitution property** of the delta function:

$$\int_{-\infty}^{\infty} f(x)\delta(x - x_0)\,dx = f(x_0). \tag{3.13}$$

Technically this relation is used to define the delta function in the theory of **generalized functions** (see for example Griffel, 1985). With $f(x) = 1$, (3.13) becomes

$$\int_{-\infty}^{\infty} \delta(x - x_0)\,dx = 1.$$

Furthermore, since $\delta(x) = 0$ for $x \neq 0$,

$$\int_{-\infty}^{x} \delta(x' - x_0)\,dx' = H(x - x_0) = \begin{cases} 0, & x < x_0, \\ 1, & x \geqslant x_0. \end{cases}$$

Figure 3.9

Thus we may informally regard $\delta(x - x_0)$ as the derivative of the unit step function $H(x - x_0)$. Thus it may be viewed as the density of the constant x_0.

Probability density of discrete random variables

Let X be a discrete random variable with $\Pr(X = 1) = 1 - \Pr(X = 0) = p$. Then the probability density of X is written

$$f_X(x) = (1 - p)\delta(x) + p\delta(x - 1).$$

This gives the correct distribution function for X because

$$F_X(x) = \Pr(X \leqslant x) = \int_{-\infty}^{x} f_X(x')\,dx' = (1 - p)H(x) + pH(x - 1)$$

$$= \begin{cases} 0, & x < 0, \\ 1 - p, & 0 \leqslant x < 1, \\ 1, & x \geqslant 1, \end{cases}$$

as is sketched in Fig. 3.9.

Similarly, the probability density of a Poisson random variable with parameter λ is given by

$$f_X(x) = e^{-\lambda} \sum_{k=0}^{\infty} \frac{\lambda^k}{k!} \delta(x - k).$$

3.9 AN APPLICATION IN NEUROBIOLOGY

In Section 3.5 we mentioned the small voltage changes which occur spontaneously at nerve–muscle junctions. Their **arrival times** were found to be well described by a Poisson point process in time. Here we are concerned with their **magnitudes**. Figure 3.10 depicts the anatomical arrangement at the nerve–muscle junction. Each cross represents a potentially active site.

The small **spontaneous** voltage changes have amplitudes whose histogram is fitted to a normal density – see Fig. 3.11. When a **nerve impulse**, having travelled out from the spinal cord, enters the junction it elicits a much bigger

Figure 3.10 The arrangement at a nerve–muscle junction.

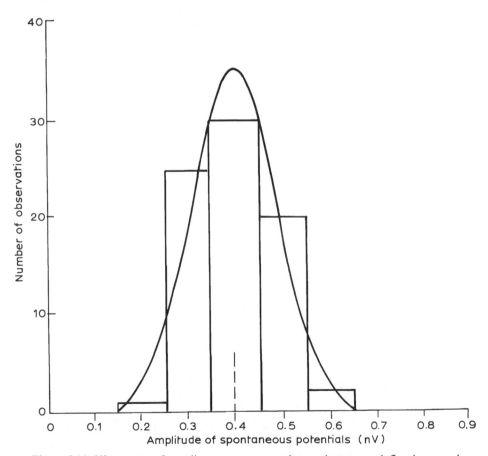

Figure 3.11 Histogram of small spontaneous voltage changes and fitted normal density. From Martin (1977). Figures 3.11–3.13 reproduced with permission of the American Physiological Society and the author.

response whose amplitude we will call V. It was hypothesized that the large response was composed of many unit responses, the latter corresponding to the spontaneous activity.

We assume that the unit responses are $X_1, X_2, \ldots$ and that these are normal with mean μ and variance σ^2. A large response consists of a random number N of the unit responses. If $N = 0$, there is no response at all. Thus

$$V = X_1 + X_2 + \cdots + X_N,$$

which is a random sum of random variables. A natural choice for N is a binomial random variable with parameters n and p where n is the number of potentially active sites and p is the probability that any site is activated. However, the assumption is usually made that N is Poisson. This is based on the Poisson approximation to the binomial and the fact that a Poisson distribution is characterized by a single parameter. Hence V is a compound Poisson random variable. The probability density of V is then found as follows:

$$\Pr\{V \in (v, v + dv)\} = \sum_{k=0}^{x} \Pr\{V \in (v, v + dv) \mid N = k\} \Pr\{N = k\}$$

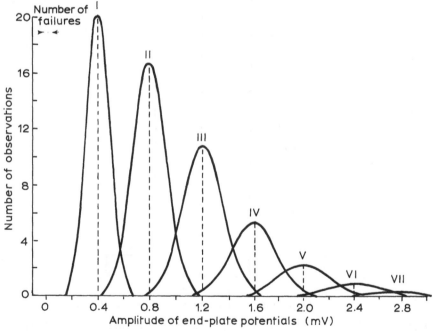

Figure 3.12 Decomposition of the compound Poisson distribution. The curve marked I corresponds to p_1, the curve marked II to p_2, etc., in (3.14).

$$= e^{-\lambda} \Pr\{V \in (v, v + dv) | N = 0\}$$

$$+ \sum_{k=1}^{\infty} \frac{e^{-\lambda}\lambda^k}{k!} \Pr\{V \in (v, v + dv) | N = k\}$$

$$= \left[e^{-\lambda}\delta(v) + e^{-\lambda} \sum_{k=1}^{\infty} \frac{\lambda^k}{k!} \frac{1}{\sqrt{2\pi k\sigma^2}} \exp\left(\frac{-(v - k\mu)^2}{2k\sigma^2} \right) \right] dv$$

$$\doteq \left[e^{-\lambda}\delta(v) + \sum_{k=1}^{\infty} p_k(v) \right] dv,$$

where $\delta(v)$ is a delta function concentrated at the origin. Hence the required density is

$$\boxed{f_V(v) = e^{-\lambda}\delta(v) + \sum_{k=1}^{\infty} p_k(v)}. \tag{3.14}$$

The terms in the expansion of the density of V are shown in Fig. 3.12. The density of V is shown in Fig. 3.13 along with the empirical distribution. Excellent agreement is found between theory and experiment, providing a validation of the 'quantum hypothesis'. For further details see Martin (1977).

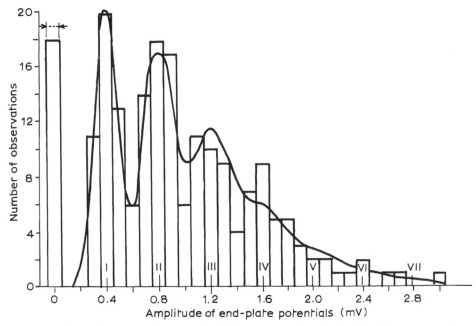

Figure 3.13 Histogram of responses. The curve is the density for the compound Poisson distribution, the column at 0 corresponding to the delta function in (3.14).

REFERENCES

Afifi, A.A. and Azen, S.P. (1979). *Statistical Analysis. A Computer Oriented Approach.* Academic Press, New York.

Begon, M. (1979). *Investigating Animal Abundance; Capture-recapture for Biologists.* Edward Arnold, London.

Blake, I.F. (1979). *An Introduction to Applied Probability.* Wiley, New York.

Clayton, G. (1984). Spatial processes with special emphasis on the use of distance methods in analysing spatial point processes. PhD Thesis, Monash University.

Cormack, R. M. (1968). The statistics of capture-recapture methods. *Oceanogr. Mar. Biol. Ann. Rev.,* **6,** 455–506.

Cormack, R.M. (1973). Commonsense estimates from capture-recapture studies. In *The Mathematical Theory of the Dynamics of Biological Populations* (eds M.S. Bartlett and R.W. Hiorns). Academic Press, New York, pp. 225-35.

Cox, D.R. and Lewis, P.A.W. (1966). *The Statistical Analysis of Series of Events.* Methuen, London.

Diggle, P.J. (1975). Robust density estimation using distance methods. *Biometrika,* **62,** 39–48.

Diggle, P.J. (1983). *Statistical Analysis of Spatial Point Processes.* Academic Press, London.

Fatt, P. and Katz, B. (1952). Spontaneous subthreshold activity at motor nerve endings. *J. Physiol.,* **117,** 109-28.

Feller, W. (1968). *An Introduction to Probability Theory and its Applications,* vol. 1. Wiley, New York.

Griffel, D.H. (1985). *Applied Functional Analysis.* Ellis Horwood, Chichester.

Heltshe, J.F. and Ritchey, T.A. (1984). Spatial pattern detection using quadrat samples. *Biometrics,* **40,** 877-85.

Hoel, P.G. (1971). *Introduction to Mathematical Statistics.* Wiley, New York.

Kendall, M.G. and Stuart, A. (1958). *The Advanced Theory of Statistics,* vol. 1, *Distribution Theory.* Hafner, New York.

Liebermann, G.J. and Owen, D.B. (1961). *Tables of the Hypergeometric Distribution.* Stanford University Press, Stanford.

MacArthur, R.H. and Connell, J.H. (1966). *The Biology of Populations.* Wiley, New York.

Manly, B.F. (1984). Obtaining confidence limits on parameters of the Jolly-Seber model for capture-recapture data. *Biometrics,* **40,** 749-58.

Martin, A.R. (1977). Junctional Transmission II. Presynaptic mechanisms. In *Handbook of Physiology. The Nervous System,* vol. 1 (ed. E.R. Kandel). Amer. Physiol. Soc., Bethesda, Md.

Moran, P.A.P. (1968). *Introduction to Probability Theory.* Clarendon Press, Oxford.

Patil, G.P., Pielou, E.C. and Waters, W.E. (eds) (1971). *Statistical Ecology,* vol. 1, *Spatial Patterns and Statistical Distributions.* Penn. State Univ. Press, University Park.

Ripley, B.D. (1981). *Spatial Statistics.* Wiley, New York.

Strickberger, M.W. (1968). *Genetics.* Macmillan, New York.

Upton, G. and Fingleton, B. (1985). *Spatial Data Analysis by Example,* vol. 1. Wiley, New York.

Van der Kloot, W., Kita, H. and Cohen, I. (1975). The timing and appearance of miniature end-plate potentials. *Prog. Neurobiol.,* **4,** 269-326.

EXERCISES

1. Show by direct calculation from

$$E(X) = \sum_k k \, \Pr(X = k)$$

that if X is a hypergeometric random variable with parameters M, n and N, then $E(X) = Mn/N$. (*Hint*: Assuming $M < n$,

$$\binom{N}{n} = \sum_{k=0}^{M} \binom{M}{k} \binom{N-M}{n-k}.$$

Factor nM/N out of the expression for $E(X)$.)

2. A lake contains an unknown number of fish. One thousand fish are caught, marked and returned to the lake. At a later time a thousand fish are caught and 100 of them have marks. Estimate the total number of fish. These are Feller's (1968) figures. He went on to say: 'figures would justify a bet that the true number of fish lies somewhere between 8500 and 12000'. Using an approximate 95% confidence interval for $\hat{N}$, assess Feller's claim.

3. Without doing any calculations deduce that the expected value of $\hat{N}$ is infinite.

4. A bank dispenses ten thousand \$1 bills and finds out later that 100 of them are counterfeit. Subsequently 100 of the \$1 bills are recovered.
   (a) Write down an exact expression for the probability p that there are at least 2 counterfeit bills among them.
   (b) Using the normal approximation (with continuity correction) show that

$$p \simeq .307.$$

5. In the inverse sampling method of capture–recapture, the sampling continues until a predetermined number, m, of marked individuals is obtained. Thus the sample size is a random variable Y. Under the approximation of sampling with replacement, show that Y has a negative binomial distribution (see p. 193).

6. (a) For inverse sampling show that the maximum likelihood estimate for N is

$$\hat{N} = \frac{MY}{m}.$$

   (b) An estimator is **unbiased** if its expectation is equal to the quantity it estimates. We saw in Exercise 3 that for direct sampling the estimator of N is biased. Show that in inverse sampling $\hat{N}$ is an unbiased estimator for N.

7. A population of size N contains initially M marked individuals but with probability q each marked individual loses its mark. A sample of size n is subsequently obtained. Under the approximation of sampling with replacement, show that the mean and variance of X, the number of marked individuals in the sample are, with $p = 1 - q$,

$$E(X) = \frac{Mnp}{N}, \qquad \text{Var}(X) = \frac{Mnp}{N}\left(1 - \frac{Mp}{N}\right).$$

8. Show that the time interval between two events in a homogeneous Poisson point process with intensity λ has an exponential distribution with mean $1/\lambda$. (*Hint:* Use the law of total probability.)

9. If N_1 and N_2 are independent homogeneous Poisson point processes (HPPP) with intensities λ_1 and λ_2, show that $N = N_1 + N_2$ is HPPP with intensity

$$\lambda = \lambda_1 + \lambda_2.$$

10. Vehicles, consisting of cars and trucks, arrive at a checkpoint as HPPP in time with intensity λ. If the probability that a given vehicle is a car is p, show that the arrival times of cars are HPPP with intensity λp.

11. Northbound cars arrive at an intersection regularly at times $a, 2a, 3a, \ldots$ whereas eastbound cars arrive at random (HPPP) with mean rate λ. If a northbound and eastbound car arrive within ε of each other, there is a collision. Assuming $\varepsilon < a/2$, show that after the arrival of n northbound cars the probability p_n of no collisions is

$$p_n = e^{-2\varepsilon\lambda n}.$$

12. Consider an HPPP with intensity λ in three dimensions. Show that the distance to the nearest point has density

$$f_R(r) = 4\lambda\pi r^2 \exp(-4\lambda\pi r^3/3).$$

13. Water in a reservoir is contaminated with bacteria which occur randomly with mean rate 1 per 10 litres. If a person drinks a litre of water every day, show that the probability he swallows one or more bacteria in one week is about .5034.

14. With the same setup as in Exercise 13, show that the probability that the distance between two bacteria is greater than 10 cm is

$$p = \exp(-2\pi/15) \simeq .66.$$

15. Let N be an HPPP on the line. Given that one event occurred in $(0, t]$ show that its time of occurrence is uniformly distributed on $(0, t)$.

16. Let N be an HPPP on the line. Show that

$$\text{Cov}\,[N_{(t_1, t_2)}, N_{(t_3, t_4)}] = \lambda |(t_1, t_2) \cap (t_3, t_4)|$$

where $|(s, t)| = t - s$ denotes the length of an interval.

17. Let an HPPP start at $t = 0$. Let $s > 0$ be an arbitrary but fixed time point. Find the density of the time interval back from s to the most recent event.

18. A finite Poisson forest with an average number λ of trees per unit area is divided into N cells of equal area A. Show that the expected number of cells containing k trees is

$$n_k^* = N \exp(-\hat{n})\hat{n}^k/k!, \qquad k = 0, 1, \ldots$$

where $\hat{n}$ is the expected number of trees per cell.

19. Using the result of Exercise 18, test whether the spatial distribution in Fig. 3.14 is a Poisson forest. (Use χ^2.)

20. If R is the distance to the nearest object in a Poisson forest with intensity λ, show that R^2 is exponentially distributed with mean $1/\lambda\pi$. Use this result in Exercise 21.

21. Let $\{X_i, i = 1, 2, \ldots, n\}$ be a random sample of point-to-plant distances. Show that under the assumption of a Poisson forest an unbiased estimator of the reciprocal of λ (where λ is the expected number of plants per unit area) is

$$\hat{\Lambda}^{-1} = \frac{\pi}{n} \sum_{i=1}^{n} X_i^2.$$

22. Using a suitable subset of the Lansing Woods picture (Fig. 3.7), test by one or more of the methods outlined in Section 3.6 (or otherwise) the Poisson forest assumption. When doing this ignore the distinction between small and large dots. Assuming the Poisson hypothesis is valid, estimate the total number of trees in the portion of the forest shown.

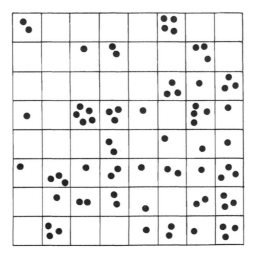

Figure 3.14

23. When a needle of length $L < 1$ is dropped onto a surface marked with parallel lines distance 1 apart, the probability of an intersection is $2L/\pi$. If when the needle lands it breaks into $N + 1$ pieces of equal length, where N is Poisson with parameter λ:

   (a) find an expression for the probability distribution of the total number of intersections;
   (b) show that the expected number of intersections is $2L/\pi$.

24. Let N be binomial with parameters n and p and let $\{X_k, k = 1, 2, \ldots\}$ be i.i.d. with mean μ and variance σ^2. Let $S_N = X_1 + X_2 + \cdots + X_N$.

   Show
   $$E[S_N] = \mu n p$$
   $$\mathrm{Var}\,[S_N] = np[\mu^2(1 - p) + \sigma^2].$$

25. Verify the substitution property

   $$\int f(x)\delta(x)\,dx = f(0),$$

   by evaluating

   $$\lim_{\varepsilon \to 0} \int f(x)\delta_\varepsilon(x)\,dx$$

   when (a) $f(x) = \cos x$, (b) $f(x) = e^x$.

4
Reliability theory

In this chapter we will be concerned with the duration of the useful life of components and systems of components. The study of this subject is called **reliability theory**. Its applications are diverse, including all forms of transportation devices, power systems, radio and television equipment, buildings and bridges, defence installations, etc. Its importance has been dramatically presented in recent disasters in nuclear power stations and space flight.

In medical science the concept of reliability is replaced by **survivability**. The time to failure becomes the time of death and this is uncertain. Of especial interest is a comparison of the survival times of patients under different courses of medical treatment. As can be anticipated, much of the theory developed for reliability theory is useful in the study of survivorship. We will not, however, emphasize the medical applications. The reader who is interested in such matters may consult, for example, Gross and Clark (1975) and Cox and Oakes (1983).

4.1 FAILURE TIME DISTRIBUTIONS

The operation of complicated pieces of machinery and mechanical or electrical equipment depends on the proper functioning of their components. Since how long a component remains functional is not usually predictable we may define for any component a non-negative random variable T which is the **time to failure** or **failure time**. We let the distribution function of T be F so

$$F(t) = \Pr\{\text{Component fails at or before time } t\}$$
$$= \Pr\{T \leq t\}, \qquad t \geq 0.$$

The **density** of T will be denoted by f. The point $t = 0$ will correspond to a convenient reference such as the time of manufacture, the time of installation or the time of first use.

The actual failure time distribution for a particular device must be estimated empirically. For the purpose of analysis, however, it is convenient to have a repertoire of analytic expressions which may be used to approximate the empirical distribution. A few commonly used ones are as follows.

Gamma density

As given earlier, but repeated here for ease of reference, the gamma density with parameters λ and ρ is

$$f(t) = \frac{\lambda(\lambda t)^{\rho - 1}}{\Gamma(\rho)} e^{-\lambda t}, \qquad t > 0; \qquad \lambda, \rho > 0;$$

where $\Gamma(\rho)$ is the gamma function

$$\Gamma(\rho) = \int_0^\infty x^{\rho - 1} e^{-x} \, dx, \qquad \rho > 0.$$

It will be shown in Exercise 1 that if $n \geqslant 1$ is an integer, then

$$\Gamma(n) = (n - 1)!$$

For the gamma density the mean and variance of the failure time are

$$E(T) = \frac{\rho}{\lambda},$$

$$\text{Var}(T) = \frac{\rho}{\lambda^2}.$$

If time is measured in units of $(1/\lambda)$ (so $t' = \lambda t$) the parameter λ can be set equal to unity. If $\rho = n$, a positive integer, the distribution is also called **Erlangian**. The special case $n = 1$ gives an exponentially distributed failure time. Gamma densities are sketched for $n = 1, 2$ and 4 in Fig. 3.4.

The distribution function is (see Exercise 3), for $\rho = n$,

$$F(t) = 1 - e^{-\lambda t} \left[1 + \lambda t + \frac{(\lambda t)^2}{2!} + \cdots + \frac{(\lambda t)^{n-1}}{(n-1)!} \right] \qquad (4.1)$$

which in the case $n = 1$ is just

$$F(t) = 1 - e^{-\lambda t}, \qquad t \geqslant 0.$$

Weibull distribution

A Weibull density with parameters λ and ρ is

$$f(t) = (\lambda \rho) t^{\rho - 1} \exp[-\lambda t^\rho], \qquad t > 0; \qquad \lambda, \rho \geqslant 0. \qquad (4.2)$$

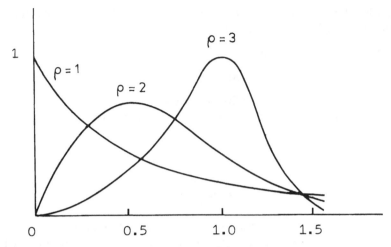

Figure 4.1 Weibull densities for $\rho = 1, 2, 3$ and $\lambda = 1$.

The mean and variance of T are shown in Exercise 5 to be

$$E(T) = \lambda^{-1/\rho} \Gamma \left(1 + \frac{1}{\rho} \right),$$

$$\mathrm{Var}(T) = \lambda^{-2/\rho} \left[\Gamma \left(1 + \frac{2}{\rho} \right) - \Gamma^2 \left(1 + \frac{1}{\rho} \right) \right].$$

Note that ρ is dimensionless and since $\lambda^{-1/\rho}$ has dimensions of time, λ has dimension of $(1/\mathrm{time})^\rho$.

When $\rho = 1$ the **exponential density** is obtained and when $\rho = 2$ the density is called a **Rayleigh density**. Weibull densities for $\rho = 1, 2$ and 3 with $\lambda = 1$ are sketched in Fig. 4.1.

The distribution function is

$$F(t) = 1 - \exp\left[-\lambda t^\rho \right] \tag{4.3}$$

It may be checked by differentiation that this gives the density (4.2).

Other failure time distributions are sometimes employed (see for example Blake, 1979). One, the truncated normal, is discussed in Exercise 6.

4.2 RELIABILITY FUNCTION AND FAILURE RATE FUNCTION

Instead of focusing on failure we may accentuate the positive and consider the following.

Definition The reliability function, $R(t)$, $t \geqslant 0$, is the probability that the component is still operating at time t.

This must be $\Pr\{T > t\}$ so

$$R(t) = 1 - F(t). \tag{4.4}$$

Another name for R is the **survivor function**.

Consider now the definition of the probability density of T:

$$f(t) = \frac{\mathrm{d}F}{\mathrm{d}t} = \lim_{\Delta t \to 0} \frac{F(t + \Delta t) - F(t)}{\Delta t}$$

$$= \lim_{\Delta t \to 0} \frac{\Pr\{t < T \leqslant t + \Delta t\}}{\Delta t}.$$

Roughly speaking, $f(t)\Delta t$ is the probability that failure occurs in $(t, t + \Delta t]$.

We now condition on the event 'failure has not yet occurred at time t', which means $T > t$.

Definition The failure rate function is

$$r(t) = \lim_{\Delta t \to 0} \frac{\Pr\{t < T \leqslant t + \Delta t \mid T > t\}}{\Delta t}. \tag{4.5}$$

Thus $r(t)\Delta t$ is approximately the probability of failure in $(t, t + \Delta t]$ given successful operation of a component to time t; i.e. it hasn't broken down at t. Sometimes $r(t)$ is called the **hazard function**.

Relationships between the various quantities defined above are contained in the following.

Theorem 4.1

(i) **The density of the failure time, the failure rate function and reliability function are related by**

$$r(t) = \frac{f(t)}{R(t)}. \tag{4.6a}$$

(ii) **In terms of the failure rate function, the reliability function is given by**

$$R(t) = \exp\left[-\int_0^t r(s)\,\mathrm{d}s \right]. \tag{4.6b}$$

Proof By definition of conditional probability

$$\Pr\{t < T \leqslant t + \Delta t \mid T > t\} = \frac{\Pr\{t < T \leqslant t + \Delta t, T > t\}}{\Pr\{T > t\}}.$$

$$= \frac{\Pr\{t < T \leqslant t + \Delta t\}}{\Pr\{T > t\}}$$

$$= \frac{F(t + \Delta t) - F(t)}{R(t)}.$$

Hence, from the definition (4.5),

$$r(t) = \lim_{\Delta t \to 0} \frac{1}{R(t)} \frac{F(t + \Delta t) - F(t)}{\Delta t}$$

which gives

$$r(t) = f(t)/R(t). \tag{4.7}$$

But from (4.4),

$$f(t) = dF/dt = \frac{d}{dt}(1 - R) = -\frac{dR}{dt}.$$

Substituting in (4.7),

$$-r(t) = \frac{1}{R}\frac{dR}{dt} = \frac{d}{dt}(\ln R), \tag{4.8}$$

Integrating from zero to t,

$$\left[\ln R(s) \right]_0^t = -\int_0^t r(s)\,ds,$$

and since $R(0) = 1$,

$$\ln R(t) = -\int_0^t r(s)\,ds.$$

This gives (4.6b) on exponentiating.

We now determine the failure rate and reliability functions for the failure time distributions discussed in Section 4.1.

Exponential distribution

Let T be exponentially distributed, so $F(t) = 1 - e^{-\lambda t}$. Then the reliability function $R = 1 - F$ is

$$R(t) = e^{-\lambda t}.$$

From (4.6a) the failure rate function is $f/R = \lambda e^{-\lambda t}/e^{-\lambda t}$ so,

$$r(t) = \lambda.$$

Thus the failure rate is constant for an exponentially distributed failure time.

An exponentially distributed failure time also has the following interesting property.

Theorem 4.2 Let T be exponentially distributed with reliability function $R(t) = e^{-\lambda t}$. Given survival to time s, the conditional survival probability for a time interval of length t beyond s is

$$\Pr\{T > s + t \mid T > s\} = e^{-\lambda t},$$

for all $s \geqslant 0$.

Proof By definition of conditional probability

$$\Pr\{T > s + t \mid T > s\} = \frac{\Pr\{T > s + t, T > s\}}{\Pr\{T > s\}}$$

$$= \frac{\Pr\{T > s + t\}}{\Pr\{T > s\}}$$

$$= \frac{e^{-\lambda(s + t)}}{e^{-\lambda s}} = e^{-\lambda t}.$$

Thus, if it is known that such a component has lasted to time s, the probabilities of future survival are the same as they were at the beginning. This is called a **memory-less property**: if the component is still operating it is 'as good as new'.

Gamma distribution

Let T have a gamma density with parameters λ and $\rho = n$, where n is a positive integer. Then, the reliability function is $R = 1 - F$ which gives, from (4.1),

$$R(t) = e^{-\lambda t}\left[1 + \lambda t + \frac{(\lambda t)^2}{2!} + \cdots + \frac{(\lambda t)^{n-1}}{(n-1)!}\right]. \qquad (4.9)$$

Since the density of T is

$$f(t) = \frac{\lambda(\lambda t)^{n-1} e^{-\lambda t}}{\Gamma(n)} = \frac{\lambda(\lambda t)^{n-1} e^{-\lambda t}}{(n-1)!},$$

the failure rate is

$$r(t) = f(t)/R(t)$$

$$= \frac{\lambda(\lambda t)^{n-1}}{(n-1)!}\left[1 + \lambda t + \frac{(\lambda t)^2}{2!} + \cdots + \frac{(\lambda t)^{n-1}}{(n-1)!}\right]^{-1}.$$

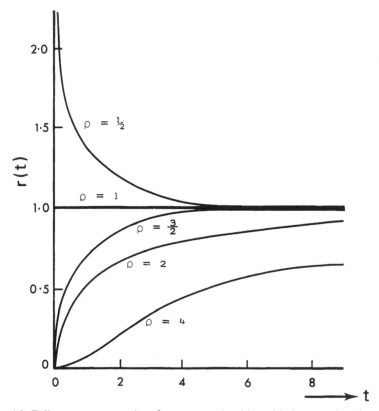

Figure 4.2 Failure rate versus time for gamma densities with $\lambda = 1$ and various ρ.

Asymptotic behaviour

It is of interest to find the long-term behaviour of the failure rate. For gamma densities a particularly simple result is obtained. As $t \to \infty$, we find

$$\lim_{t \to \infty} r(t) = \lambda$$

the approach to λ being from above or below according to the cases $\rho < 0$ and $\rho > 0$ respectively (see Exercise 7). Failure rate functions for gamma densities with various parameters are shown in Fig. 4.2.

Weibull distribution

If T has a Weibull density then, from (4.3) and the fact that $R = 1 - F$, we find the reliability function is

$$R(t) = \exp[-\lambda t^{\rho}].$$

Also, the density of T is

$$f(t) = (\lambda \rho)t^{\rho - 1} \exp(-\lambda t^{\rho})$$

so we get

$$r(t) = f(t)/R(t)$$
$$= (\lambda \rho)t^{\rho - 1}.$$

Asymptotic behaviour

As shown in Fig. 4.3 the failure rate is a decreasing function of t if $\rho < 1$ and an increasing function of t if $\rho > 1$. In contrast to the gamma case, $r(t)$ does not asymptote to the constant λ but rather becomes infinite ($\rho > 1$) or approaches zero ($\rho < 1$).

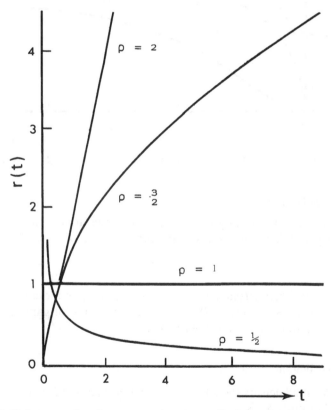

Figure 4.3 Failure rate functions versus time for Weibull distributions with $\lambda = 1$ and various ρ.

4.3 THE SPARE PARTS PROBLEM

Consider a component with failure time T_1. Since it is known that the component may fail, spares are carried. The important question arises as to how many spares must be carried to guarantee a specified probability of continuous operation for a given length of time. It is assumed that the original and spares have the same failure time distribution with density f, though this is not necessary.

One spare

Let T_1 be the time at which the original fails and T_2 that at which the spare fails after its installation. Continuous operation (assuming instantaneous replacement) to time t occurs if

$$T = T_1 + T_2 > t.$$

From Section 2.4, the sum of two independent random variables has a probability density which is the convolution of the densities of the separate random variables. Thus the density of T is

$$f_T(t) = \int_0^t f(t - t')f(t')\,dt', \qquad t > 0.$$

The requirement that the probability of continuous operation in $(0, t)$ be at least α gives

$$\Pr\{T > t\} = \int_t^\infty f_T(s)\,ds = R_T(t) \geqslant \alpha.$$

Several spares

If there is an original and $n - 1$ spares, the effective failure time is

$$T = T_1 + T_2 + \cdots + T_n$$

whose density is an **n-fold convolution** of the density of T_1:

$$f_T(t) = \int_0^t \int_0^{t_1} \cdots \int_0^{t_{n-1}} f(t - t_1)f(t_1 - t_2)\cdots f(t_{n-1} - t_n)f(t_n)\,dt_n\,dt_{n-1}\cdots dt_1.$$

Exponentially distributed failure times

If there are parts numbered $1, 2, \ldots, n$, and each has an exponential distribution with mean $1/\lambda$, then as we may infer from the results of Chapter 3, $T = \sum_{i=1}^n T_i$ has a gamma distribution with parameters λ and n. Then, from (4.9), the minimum number of parts required for continuous operation to time

t with probability at least α is the smallest n such that

$$R_T(t) = e^{-\lambda t}\left[1 + \lambda t + \frac{(\lambda t)^2}{2!} + \cdots + \frac{(\lambda t)^{n-1}}{(n-1)!}\right] \geq \alpha. \qquad (4.10)$$

Note that as $n \to \infty, R_T(t) \to 1$.

Example A component of an electronic circuit has an exponentially distributed failure time with mean 1000 hours. What is the smallest number of spares which must be carried to guarantee continuous operation for 1000 hours with probability at least .95?

Solution If we work with time units of 1000 hours we may set $\lambda = 1$. The number of components needed is the smallest n satisfying

$$R_T(1) = e^{-1}\left[1 + 1 + \frac{1}{2!} + \cdots + \frac{1}{(n-1)!}\right] \geq .95$$

where (4.10) is used with $\lambda = t = 1$. We find

$$e^{-1}\left[1 + 1 + \frac{1}{2!}\right] \simeq .920$$

$$e^{-1}\left[1 + 1 + \frac{1}{2!} + \frac{1}{3!}\right] \simeq .981.$$

Hence $n - 1 = 3$, so the smallest number of components is four; that is, an original plus three spares.

Failures as a Poisson process

The assumption of independent identically exponentially distributed time intervals between the failures of the various parts is equivalent to saying that the failures occur as events in a homogeneous Poisson point process (see Section 3.4). This observation leads to the generalization of the above result to several components (see Exercise 12).

4.4 COMPLEX SYSTEMS

A complex system is defined as one whose operation depends on the integrity of more than one component.

Two components

We consider a system of two components with failure times T_1 and T_2, with corresponding densities f_i, distribution functions F_i and reliability functions

R_i, $i = 1, 2$. Regardless of the connections between the two components we ask, 'What is the probability that a particular one of the components will fail before the other?' or, equivalently, 'What is the probability that one of the components will need to be replaced before the other?' If component 1 fails first then we have observed the event $\{T_1 < T_2\}$. If the components act independently, the following result gives the required probability.

Theorem 4.3

$$\Pr\{T_1 < T_2\} = \int_0^\infty F_1(t) f_2(t)\, dt. \tag{4.11}$$

Proof Since it is assumed that the components act stochastically independently of each other, the joint density of T_1, T_2 is the product of their densities. The required probability is obtained by integrating the joint density over the shaded region in Fig. 4.4.

Hence

$$\Pr\{T_1 < T_2\} = \int_{t_2 = 0}^\infty \int_{t_1 = 0}^{t_2} f_1(t_1) f_2(t_2)\, dt_1\, dt_2$$

$$= \int_{t_2 = 0}^\infty \left[F_1(t_2) - F_1(0) \right] f_2(t_2)\, dt_2 \tag{4.12}$$

$$= \int_{t_2 = 0}^\infty F_1(t_2) f_2(t_2)\, dt_2.$$

A closed-form expression for (4.11) can be obtained in some cases. For example, if T_1, T_2 are gamma distributed with parameters n_1, λ_1 and n_2, λ_2,

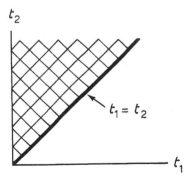

Figure 4.4 Region of integration in (4.12).

respectively, where n_1 and n_2 are positive integers, then

$$\Pr\{T_1 < T_2\} = \frac{(1 - \lambda)^{n_1}}{\lambda(n_1 - 1)!} \sum_{k=1}^{n_2} \frac{\lambda^k(n_1 + k - 2)!}{(k - 1)!} \qquad (4.13)$$

$$\lambda = \frac{\lambda_2}{\lambda_1 + \lambda_2}.$$

Two special cases are of interest. If $n_1 = n_2 = 1$, so both components have exponentially distributed failure times, then

$$\Pr\{T_1 < T_2\} = \frac{\lambda_1}{\lambda_1 + \lambda_2} \qquad (4.14)$$

If $\lambda_1 = \lambda_2$, then $\lambda = \frac{1}{2}$ and (4.14) becomes

$$\Pr\{T_1 < T_2\} = \frac{2^{1-n_1}}{(n_1 - 1)!} \sum_{k=1}^{n_2} \frac{2^{-k}(n_1 + k - 2)!}{(k - 1)!} \qquad (4.15)$$

Formulas (4.13)–(4.15) are established in Exercise 13.

4.5 SERIES AND PARALLEL SYSTEMS

The failure time distribution of a complex system depends on the interrelationships between its parts. In this section and the next we investigate this

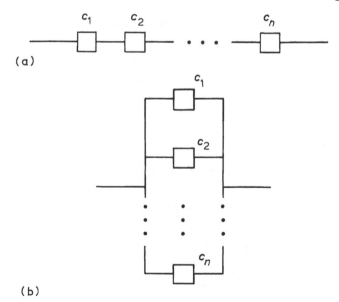

(a)

(b)

Figure 4.5 A series arrangement (a) and a parallel arrangement (b) of n components.

dependence in several interesting cases. In Fig. 4.5a a **series arrangement** of n components, C_i, $i = 1, 2, \ldots, n$ is shown, and in Fig. 4.5b a **parallel arrangement** is shown.

The following assumptions are made:

(A) In a series arrangement the system operates only if *all* the components are operating: it fails if one or more components fail.
(B) In a parallel arrangement the system operates if at least one of the components is operating: it fails only if all of its components fail.

Let T be the failure time of the whole system. Suppose each component acts independently of all the others. Denote by F_k, R_k, r_k the failure time distribution, reliability function and failure rate of the kth component. Then we have the following result.

Theorem 4.4 For n components in series, the system reliability function is

$$R(t) = \prod_{k=1}^{n} R_k(t), \qquad (4.16)$$

and for n components in parallel, the system failure time distribution is

$$F(t) = \prod_{k=1}^{n} F_k(t). \qquad (4.17)$$

Proof For the series system

$$R(t) = \Pr\{T > t\} = \Pr\{\text{system operating at } t\}$$
$$= \Pr\{\text{all components operating at } t\}.$$

By independence this is the product of the probabilities that each component is still operating at t.

For the parallel system

$$F(t) = \Pr\{T \leqslant t\} = \Pr\{\text{system fails at or before } t\}$$
$$= \Pr\{\text{all components fail at or before } t\}.$$

Again by independence the result follows.

The following results are also useful.

Corollaries to Theorem 4.4

(i) For n components in series

(a) $r(t) = \displaystyle\sum_{k=1}^{n} r_k(t)$

(b) If each component has an exponentially distributed failure time and $F_k(t) = 1 - \exp(-\lambda_k t)$, then the whole system has an exponentially distributed failure time with $F(t) = 1 - \exp(-\lambda t)$ where $\lambda = \lambda_1 + \lambda_2 + \cdots + \lambda_n$.

Hence the failure rate is $r(t) = \lambda$.

(ii) For n components in parallel,

(a) $R(t) = 1 - \prod_{k=1}^{n} (1 - R_k(t))$.

(b) If each component has the same failure time distribution with reliability $R_1(t)$, then

$$R(t) = 1 - (1 - R_1(t))^n.$$

Furthermore, if at least m of the n components must be operating for the system to operate

$$R(t) = \sum_{k=m}^{n} \binom{n}{k} R_1^k(t)(1 - R_1(t))^{n-k}.$$

Proof

(i) (a) Using (4.8),

$$r(t) = -\frac{d}{dt} \ln R(t)$$

$$= -\frac{d}{dt} \ln \prod_{k=1}^{n} R_k(t)$$

$$= -\frac{d}{dt} \sum_{k=1}^{n} \ln R_k(t)$$

$$= \sum_{k=1}^{n} r_k(t),$$

as required.

(b) If $R_k(t) = e^{-\lambda_k t}$, then

$$R(t) = \prod_{k=1}^{n} e^{-\lambda_k t}$$

$$= \exp\left[-\left(\sum_{1}^{n} \lambda_k \right) t \right]$$

The result follows.

(ii) (a) Substitute $R(t) = 1 - F(t)$ and $F_k(t) = 1 - R_k(t)$ in (4.17).

(b) The first part is obvious. For the second part, notice that the number of successfully operating components is a binomial random variable with parameters n and $p = R_1(t)$. The given result is the probability that such a random variable is at least m.

4.6 COMBINATIONS AND OTHER STRUCTURES

In practical situations very complicated interconnections and interdependencies may occur between large numbers of components. However, complex systems can often be analysed quite efficiently. For example, the system reliability function can be readily determined for combinations of series and/or parallel arrangements.

Example 1

Consider the system shown in Fig. 4.6a where elements C_1, C_2 in series are in parallel with C_3. The system is clearly equivalent to the one in Fig. 4.6b where C_4 is a component with the reliability of the C_1, C_2 series system. Since, from (4.16)

$$R_4 = R_1 R_2$$

and from Corollary (ii) (a) of Theorem 4.4,

$$R = 1 - (1 - R_3)(1 - R_4),$$

the system reliability function is

$$R = 1 - (1 - R_1 R_2)(1 - R_3).$$

Example 2

Figure 4.7a shows a **bridge structure**. The system is operating if at least one of the following series arrangements is operating: $C_1 C_2$; $C_3 C_4$; $C_1 C_5 C_4$; or $C_3 C_5 C_2$. Hence the system is equivalent to that shown in Fig. 4.7b. Denoting the event {Component C_i is operating at time t} by A_i, $i = 1, \ldots, 5$, we see that

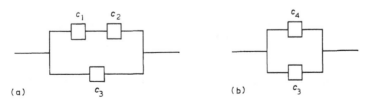

(a) c_3 (b) c_3

Figure 4.6

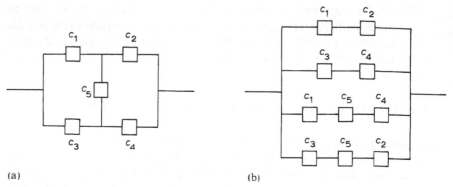

(a)   (b)

Figure 4.7

the reliability function of the bridge structure is

$$R(t) = P\{(A_1A_2 \cup A_3A_4) \cup (A_1A_4A_5 \cup A_2A_3A_5)\}. \tag{4.18}$$

To find $R(t)$ in terms of the reliability functions of the components we define

$$A = A_1A_2$$
$$B = A_3A_4$$
$$C = A_1A_4A_5$$
$$D = A_2A_3A_5$$
$$E = A \cup B$$
$$F = C \cup D.$$

The required probability can now be written as the probability of the union of the events E and F. From Chapter 1 we have the basic formula

$$P\{E \cup F\} = P\{E\} + P\{F\} - P\{EF\}. \tag{4.19}$$

Repeated use of this formula gives

$$P\{E\} = P\{A_1A_2\} + P\{A_3A_4\} - P\{A_1A_2A_3A_4\}$$

and

$$P\{F\} = P\{A_1A_4A_5\} + P\{A_2A_3A_5\} - P\{A_1A_2A_3A_4A_5\}.$$

To complete the calculation we need the following formula which is established in Exercise 18.

$$P\{(A \cup B)(C \cup D)\} = P\{AC\} + P\{AD\} + P\{BC\} + P\{BD\}$$
$$- P\{ABD\} - P\{ACD\} - P\{BCD\} + P\{ABCD\}. \tag{4.20}$$

Using this formula we find the contribution from $P\{EF\}$ in (4.19) is

$$P\{EF\} = P\{A_1A_2A_4A_5\} + P\{A_1A_3A_4A_5\} + P\{A_1A_2A_3A_5\}$$
$$+ P\{A_2A_3A_4A_5\} - 3P\{A_1A_2A_3A_4A_5\}.$$

Substituting in (4.19) and using the assumed independence of each component we find (omitting the argument t throughout)

$$R = R_1 R_2 + R_3 R_4 + R_1 R_4 R_5 + R_2 R_3 R_5 - R_1 R_2 R_3 R_4$$
$$- R_1 R_2 R_3 R_5 - R_1 R_2 R_4 R_5 - R_1 R_3 R_4 R_5 - R_2 R_3 R_4 R_5$$
$$+ 2 R_1 R_2 R_3 R_4 R_5. \tag{4.21}$$

In the event that each component has the same reliability $\bar{R}$ we find

$$R = \bar{R}^2 (2 + 2\bar{R} - 5\bar{R}^2 + 2\bar{R}^3).$$

FURTHER READING

For a more advanced treatment of reliability theory see the monograph by Barlow and Proschan (1975) and Chapter 9 of Ross (1985). One branch of reliability theory is **renewal theory** in which attention is focused on successive replacements of a given component. The numbers of replacements in various time intervals constitute a **renewal process**. The classic treatment of renewal theory is that of Cox (1962).

REFERENCES

Barlow, R.F. and Proschan, F. (1975). *Statistical Theory of Reliability and Life Testing*. Holt, Rinehart and Winston, New York.
Blake, I.F. (1979). *An Introduction to Applied Probability*. Wiley, New York.
Cox, D.R. (1962). *Renewal Theory*. Methuen, London.
Cox, D.R. and Oakes, D. (1983). *Analysis of Survival Data*. Chapman and Hall, London.
Gross, A.J. and Clark, V.A. (1975). *Survival Distributions: Reliability Applications in the Biomedical Sciences*. Wiley, New York.
Ross, S.M. (1985). *Introduction to Probability Models*. Academic Press, New York.

EXERCISES

1. Show that if $n \geqslant 1$ is a positive integer, then $\Gamma(n) = (n-1)!$ (*Hint*: First show $\Gamma(n) = (n-1)\Gamma(n-1)$.)

2. Show that the mean and variance of a random variable T with a gamma density are

$$E(T) = \rho/\lambda, \quad \mathrm{Var}(T) = \rho/\lambda^2.$$

3. Prove that the distribution function of a random variable which is gamma distributed with parameters λ and $\rho = n \geqslant 1$ is

$$F(t) = 1 - e^{-\lambda t} \left[1 + \lambda t + \frac{(\lambda t)^2}{2!} + \cdots + \frac{(\lambda t)^{n-1}}{(n-1)!} \right].$$

4. With time in hours a machine has a failure time which has a Weibull distribution with $\rho = .5$ and $\lambda = .1$ hours$^{-.5}$.

(a) What are the mean and variance of the failure time?
(b) What is the expectation and variance of the number of failures in the first 50 hours in 100 machines?
  (*Ans*: (a) 200 hours, 200 00 hours^2. (b) 50.7, 25.0.)

5. If T has a Weibull distribution with parameters λ and ρ, prove

$$E(T) = \lambda^{-1/\rho}\Gamma(1 + 1/\rho)$$
$$\mathrm{Var}(T) = \lambda^{-2/\rho}[\Gamma(1 + 2/\rho) - \Gamma^2(1 + 1/\rho)].$$

6. Let T have a **truncated normal density**

$$f(t) = \frac{k}{\sqrt{2\pi\sigma^2}}\exp\left[-\frac{(t-\mu)^2}{2\sigma^2}\right], \qquad t > 0.$$

Find a formula for k.

7. Let T be gamma distributed with parameters ρ and $\lambda = 1$.

(a) Show that

$$\frac{1}{r(t)} = \int_0^\infty \left(1 + \frac{u}{t}\right)^{\rho-1} e^{-u}\,du.$$

(b) Hence deduce that the failure rate is a decreasing function of t if $\rho < 1$ and an increasing function of t if $\rho > 1$. Show also that $r(\infty) = 1$.

8. Prove that if T_1, T_2 are independent and gamma distributed with parameters λ, ρ_1 and λ, ρ_2, then $T = T_1 + T_2$ is gamma distributed with parameters λ, $\rho = \rho_1 + \rho_2$.
  (*Hint*: Use the relation between the beta function

$$\beta(x, y) = \int_0^1 t^{x-1}(1 - t)^{y-1}\,dt, \qquad x > 0, \ y > 0,$$

and the gamma function

$$\beta(x, y) = \frac{\Gamma(x)\Gamma(y)}{\Gamma(x + y)}.)$$

Give an easy proof when $\rho_1 = n_1$, $\rho_2 = n_2$ are positive integers.

9. Prove that if the failure rate is constant then the failure time is exponentially distributed.

10. A machine component has an exponentially distributed failure time with a mean of 20 hours.

(i) If there are an original and just one spare, find the probability that the component is operational for at least 24 hours.

(ii) How many spares are needed to ensure continuous operation for 24 hours with probability at .95?
(*Ans*: .663, 3.)

11. This problem shows the crucial importance of back-up systems. The main power supply at a hospital has a failure time which is gamma distributed with $\lambda = 1/1000$ hours and $\rho = 4$. A standby power supply has an exponentially distributed failure time with mean 1000 hours. Find

(i) the reliability $R_1(t)$ of the main power supply;
(ii) the reliability $R(t)$ of the power system.
(iii) Compute $R_1(1000)$ and $R(1000)$.
(*Hint*: Use the result of Exercise 8.) (*Ans*: part (iii): .981, .996.)

12. A system has m locations at which the same component is used, each component having an exponentially distributed failure time with mean $1/\lambda$. When a component at any location fails it is replaced by a spare. In a given mission the component at location k is required to operate for time t_k. Show that the number of components required to ensure a successful mission with probability at least α is the smallest n satisfying

$$\sum_{j=0}^{n} \frac{e^{-\mu} \mu^j}{j!} \geq \alpha$$

where $\mu = \lambda \sum_{k=1}^{m} t_k$. (*Hint*: The number of failures at each location is a Poisson process. Use the result from Chapter 3, Exercise 9.)

13. If T_1, T_2 are independent and gamma distributed with parameters λ_1, n_1 and λ_2, n_2 prove that (c.f. (4.13))

$$\Pr\{T_1 < T_2\} = \frac{(1-\lambda)^{n_1}}{\lambda(n_1-1)!} \sum_{k=1}^{n_2} \frac{\lambda^k(n_1+k-2)!}{(k-1)!}$$

In particular, if $n_1 = n_2 = 1$,

$$\Pr\{T_1 < T_2\} = \frac{\lambda_1}{\lambda_1 + \lambda_2}.$$

14. After a person reaches age 50 the failure times of his heart and liver are gamma distributed with means of 10 years. For the heart $\rho = 3$, whereas for the liver $\rho = 2$. Assuming the heart and liver fail independently, what is the probability of heart failure before liver failure?
(*Ans*: .156.)

15. For n independent components in series deduce that the failure time is

$$T = \min(T_1, T_2, \ldots, T_n)$$

whereas for n components in parallel

$$T = \max(T_1, T_2, \ldots, T_n).$$

Figure 4.8

Hence deduce that formulas for the distribution functions of the minimum and maximum of n independent random variables are

$$F_{\min(T_1,\dots,T_n)} = 1 - \prod_{k=1}^{n} (1 - F_k)$$

$$F_{\max(T_1,T_2,\dots,T_n)} = \prod_{k=1}^{n} F_k.$$

16. Show that for n components in parallel,

$$r(t) = \sum_{k=1}^{n} \frac{r_k(t) R_k(t)}{F_k(t)}.$$

17. A plane has four engines whose failure times are gamma distributed with mean 100 hours and $\rho = 4$. If at least three engines must operate for the plane to fly, what is the probability that a flight of duration 20 hours will fail?

(*Ans*: .0005.)

18. (a) Verify formula (4.20).

(b) Use the following general result for sets A_i, $i = 1, 2, \dots, n$,

$$P\left(\bigcup_{i=1}^{n} A_i\right) = \sum_{i=1}^{n} P(A_i) - \sum_{\substack{\text{distinct} \\ \text{pairs}}} P(A_i \cap A_j)$$

$$+ \sum_{\substack{\text{distinct} \\ \text{triples}}} P(A_i \cap A_j \cap A_k)$$

$$- \cdots + (-1)^{n-1} P(A_1 \cap A_2 \cap \cdots \cap A_n)$$

to derive (4.21) from (4.18).

19. Find expressions for the reliability functions of the systems in Fig. 4.8.

5

Simulation and random numbers

5.1 THE NEED FOR SIMULATION

In many physical problems involving *random phenomena* we are unable to find exact expressions (or sometimes even expressions) for quantities of interest. We may then turn to *simulation* in which the situation of interest is reproduced theoretically and the results analysed. Simulation, when performed accurately, can be a powerful method of approximate solution for very complex problems.

An example from queueing theory

Consider the following simple queueing problem. With a view to improving its service, the management of a bank is interested in investigating the factors which determine the lengths of waiting times experienced by its customers. Suppose that the arrival times of customers of the bank are approximately a

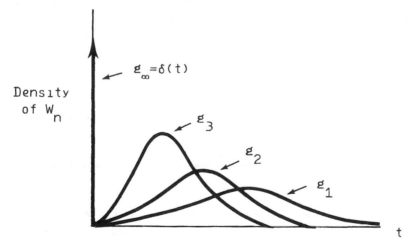

Figure 5.1 Sketch of densities of the waiting times when there are n servers, $n = 1, 2, 3, \ldots$.

Poisson point process with mean rate λ. There are $n \geqslant 1$ servers and the customer service time is a random variable S with density $f(t)$. Let W_n be the (random) time a customer has to ·wait to be served and let $g_n(t)$ be its probability density. We expect that as $n \to \infty$ the customers will always be served immediately, so

$$\lim_{n \to \infty} g_n(t) = \delta(t).$$

For $n = 1, 2, 3$ the probability densities will look something like those sketched in Fig. 5.1.

Knowledge of $g_n(t)$ for various n is useful to the operators of the bank for the following reasons.

(a) If n is too small, g_n will be more concentrated at large t which means that a lot of customers will have a long time to wait for service. This will make them impatient and they are likely to do their banking elsewhere.

(b) If n is too large, g_n will be more concentrated at small t. In this case customers will usually get rapid service but the servers will be idle for much of the time, so there will be a needless expense of paying them to do nothing. In addition, they may get bored and look for another job.

The determination of exact expressions for $g_n(t)$ is very difficult. **Computer simulation**, however, can provide some relatively quick and inexpensive estimates for g_n. A basic concept needed in simulation, that of a random sample of size n for a given random variable, was given in Section 1.3.

To estimate $g_n(t)$, $t \geqslant 0$, in the above problem by computer simulation a random sample for T, the time between customer arrivals, and a random sample for S, the service times are needed. Note that T is exponentially distributed with mean $1/\lambda$. Let the sample values of S and T be $s_1, s_2, \ldots$ and $t_1, t_2, \ldots$. Then Fig. 5.2 illustrates how, in the case of just one server, values $w_1, w_2, \ldots$ of the waiting time W may be observed in a realization of a random experiment which simulates the above service situation.

Figure 5.2 Illustrating how a random sample for S and a random sample for T may be used to obtain a sample for the customer waiting time W by simulation.

With a sufficient number of observations on the waiting times of customers, these may be collected into a histogram. The histogram will provide an estimate of the waiting time density. Similarly, the times for which the server is idle may also be collected into a histogram.

It is obvious that to perform an accurate simulation, reliable methods are needed to generate random samples for the various random variables involved. The numerical values of the random sample in the context of simulation are often referred to as **random numbers**. These are usually generated by deterministic methods called **algorithms** and are then referred to as **pseudo-random numbers**. Usually we just refer to them as random numbers and programs that produce them are called **random number generators**. In the next sections we describe some methods for generating them.

5.2 THE USEFULNESS OF A RANDOM SAMPLE FROM A UNIFORM DISTRIBUTION

We are going to show that if a random sample is available for a random variable which is uniformly distributed on (0, 1), then it may be used to obtain random samples for many other random variables.

Recall that if U is uniformly distributed on (0, 1), then its distribution function is

$$F_U(u) = \Pr\{U \leqslant u\} = \begin{cases} 0, & u \leqslant 0, \\ u, & 0 < u < 1, \\ 1, & u \geqslant 1. \end{cases}$$

Now let X be a random variable taking values in (a, b). Assume that the distribution function F, of X, is

(i) continuous on (a, b), and
(ii) strictly increasing on (a, b). That is, if $x_2 > x_1$ then $F(x_2) > F(x_1)$ for any $x_1, x_2 \in (a, b)$.

Under conditions (i) and (ii), F is guaranteed to have an inverse function denoted by F^{-1} (see introductory calculus texts). We note the following properties of inverse function:

$$F(F^{-1}(x)) = F^{-1}(F(x)) = x.$$

For example, $\ln e^x = e^{\ln x} = x$. We now prove the following result.

Theorem 5.1 **Under the above assumptions**

$$F(X) \overset{d}{=} U(0, 1),$$

where $\overset{d}{=}$ means 'has the distribution of' and $U(0, 1)$ is uniform on (0, 1).

Corollary *If* $\{U_i, i = 1, 2, \ldots\}$ *is a random sample for* U, *then*

$$\{X_i = F^{-1}(U_i), \quad i = 1, 2, \ldots\}$$

is a random sample for X.

Proof

$$\Pr\{F(X) \leqslant x\} = \Pr\{F^{-1}(F(X)) \leqslant F^{-1}(x)\}$$
$$= \Pr\{X \leqslant F^{-1}(x)\}$$
$$= F(F^{-1}(x))$$
$$= x,$$

so $F(X)$ is uniformly distributed. Furthermore, $F(a) = 0$ and $F(b) = 1$, so $F(X)$ takes values in $(0, 1)$. That is, $F(X) \overset{d}{=} U(0, 1)$. Since this is true, $X \overset{d}{=} F^{-1}(U)$ and the corollary follows.

Note that $F(X)$ is uniformly distributed on $(0, 1)$ whenever $F(x)$ is continuous, as proved in Exercise 1. We make the additional requirement that F have an inverse. The relation $F(X) = U$ is often called the **probability integral transformation**.

Example 1. Exponential distribution

It is required to generate a random sample for X where X has the distribution function

$$F(x) = 1 - e^{-\lambda x} = y, \quad \text{say.}$$

To obtain the inverse function we must express y in terms of x. We have

$$e^{-\lambda x} = 1 - y$$

so on taking logarithms and rearranging,

$$x = -\frac{1}{\lambda} \ln(1 - y) = F^{-1}(y).$$

Hence

$$X_i = F^{-1}(U_i) = -\frac{1}{\lambda} \ln(1 - U_i) \tag{5.1}$$

provides the formula for getting values of X_i from the values of U_i. For example, with $\lambda = 1$, if the first three values of U are $u_1 = .13$, $u_2 = .97$ and $u_3 = .20$; then from (5.1) the first three values for X are $x_1 = .139, x_2 = 3.507$ and $x_3 = .223$.

Example 2. Poisson distribution

Here X is a discrete random variable whose distribution function F is not continuous and hence does not meet the requirements of the above theorem.

However, we utilize the fact that the probability that a uniform $(0, 1)$ random variable takes values in any subinterval of $(0, 1)$ is just the length of that subinterval.

We have from the definition of Poisson random variable

$$\Pr\{X = k\} = e^{-\lambda}\lambda^k/k!, \qquad k = 0, 1, 2, \ldots$$

We first observe that if U is uniformly distributed on $(0, 1)$, then

$$\Pr\{0 < U < e^{-\lambda}\} = e^{-\lambda} = \Pr\{X = 0\}.$$

Similarly,

$$\Pr\{e^{-\lambda} < U < e^{-\lambda} + \lambda e^{-\lambda}\} = \lambda e^{-\lambda} = \Pr\{X = 1\}.$$

and

$$\Pr\{e^{-\lambda} + \lambda e^{-\lambda} < U < e^{-\lambda} + \lambda e^{-\lambda} + \lambda^2 e^{-\lambda}/2!\} = \lambda^2 e^{-\lambda}/2! = \Pr\{X = 2\}.$$

In fact, in general,

$$\Pr\left\{e^{-\lambda}\sum_{k=0}^{n-1}\frac{\lambda^k}{k!} < U < e^{-\lambda}\sum_{k=0}^{n}\frac{\lambda^k}{k!}\right\} = \Pr\{X = n\}, \qquad n \geqslant 1.$$

Thus a random sample for a Poisson random variable X with parameter λ can be obtained from that for a $U(0, 1)$ random variable by putting

$$X_i = \begin{cases} 0, & \text{if } U_i \in (0, e^{-\lambda}), \\ n, & \text{if } U_i \in \left(e^{-\lambda}\sum_{k=0}^{n-1}\frac{\lambda^k}{k!}, e^{-\lambda}\sum_{k=0}^{n}\frac{\lambda^k}{k!}\right), \quad n \geqslant 1. \end{cases}$$

For example, when $\lambda = 1$, if $u_1 = .13$, $u_2 = .97$ and $u_3 = .20$, then the corresponding values of X are $x_1 = 0, x_2 = 3$ and $x_3 = 0$.

The same method can be applied to any discrete random variable, X, taking on values $x_1 < x_2 < x_3 < \cdots$ with probabilities

$$\Pr\{X = x_k\} = p_k, \qquad k = 1, 2, \ldots$$

It will be seen that one then should put

$$X_i = \begin{cases} x_1, & \text{if } U_i \in (0, p_1), \\ x_n, & \text{if } U_i \in \left(\sum_{k=1}^{n-1} p_k, \sum_{k=1}^{n} p_k\right), \quad n \geqslant 2. \end{cases}$$

Unfortunately, in the case of a normal random variable the distribution function

$$F(x) = \Pr\{X < x\} = \frac{1}{\sqrt{2\pi\sigma^2}}\int_{-\infty}^{x} \exp\left[\frac{-(y-\mu)^2}{2\sigma^2}\right] dy$$

does not have an inverse which can be expressed in closed form. A method for obtaining a random sample for a normal random variable will be given in Section 5.4.

5.3 GENERATION OF UNIFORM (0, 1) RANDOM NUMBERS

It is required to generate a sequence of numbers, $\{u_i, i = 1, 2, \ldots\}$ which looks as if it is a realization of a sequence $\{U_i, i = 1, 2, \ldots\}$, of independent random variables each being uniformly distributed on $(0, 1)$. One method is the following, due to Lehmer (1951).

The linear congruential method

In this method, a sequence of non-negative integers $\{N_k, k = 1, 2, \ldots\}$ is first obtained using the relation

$$\boxed{N_{k+1} = (lN_k + m) \bmod n}, \qquad k = 1, 2, \ldots \tag{5.2}$$

where l, m and n are given integers, the first member of the sequence, N_1, also being given. (Note that the N_k are not random variables.)

The prescription (5.2) means that N_{k+1} is equal to the remainder after dividing $lN_k + m$ by n. Note that two integers are called **congruent modulo n** if their difference is an integral multiple of n. For example, 7 and 10 are congruent modulo 3, or $7 \equiv 10 \,(\bmod\, 3)$. This terminology is part of the basis for the name of the present method.

We always require

$$n > 0, \qquad 0 \leqslant N_1 < n,$$

which guarantees that the sequence $\{N_k\}$ is positive and that the first term is less than n.

All numbers in the sequence generated by the linear congruential relation (5.2) are between 0 and $n - 1$ inclusively. Hence to obtain a sequence $\{U_i, i = 1, 2, \ldots\}$ with values in $(0, 1)$, we set

$$u_k = N_k/n, \qquad k = 1, 2, \ldots$$

Example 1

Let $l = 6, m = 7, n = 5$ and use the starting value $N_1 = 2$. Then

$$N_2 = (12 + 7) \bmod 5 = 4$$
$$N_3 = (24 + 7) \bmod 5 = 1$$
$$N_4 = (6 \;+ 7) \bmod 5 = 3$$
$$N_5 = (18 + 7) \bmod 5 = 0$$
$$N_6 = (0 \;+ 7) \bmod 5 = N_1.$$

The sequence generated in $(0, 1)$ is, on dividing by 5, $\{.4, .8, .2, .6, 0, \ldots\}$.

It can be seen that the above sequence $\{N_k\}$ repeats after N_5,

$$\{2, 4, 1, 3, 0, 2, 4, 1, 3, 0, \ldots\},$$

so the sequence has **period 5**. In general, if

$$N_{p+k} = N_k, \qquad k \geqslant q \geqslant 1,$$

for some q, then the sequence eventually has period p (see also Example 2 below). Since there are only n possible distinct numbers in the sequence $\{N_k\}$, the maximum possible period is n.

Choice of parameters

In the above example with $n = 5$ the maximum period was attained. Consider now the following examples.

Example 2

The sequence generated when $N_1 = 64, l = 30, m = 60$ and $n = 100$ is

$$N_2 = (30 \times 64 + 60) \bmod 100 = (1920 + 60) \bmod 100 = 80$$
$$N_3 = (30 \times 80 + 60) \bmod 100 = (2400 + 60) \bmod 100 = 60$$
$$N_4 = (30 \times 60 + 60) \bmod 100 = (1800 + 60) \bmod 100 = 60.$$

It follows that $N_k = 60$ for all $k \geqslant 3$ and since $N_{k+1} = N_k$ for $k \geqslant 3$, the sequence ends up having period 1. This is way short of the maximum possible period of 100.

Example 3

When $l = m = 1$, the sequence is

$$N_{k+1} = (N_k + 1) \bmod n,$$

with period n. For example, if $n = 10$ and $N_1 = 3$ we have

$$\{3, 4, 5, 6, 7, 8, 9, 0, 1, 2, 3, \ldots\}$$

with period 10. Thus, even though the maximum possible period is attained we would hardly wish to call it a random sequence.

These examples illustrate that in the linear congruential method the choice of parameters is a very important consideration. We seek a sequence with a large period, much larger than the sample size required, but which looks random. The following result gives necessary and sufficient conditions for obtaining the maximum period.

Theorem 5.2 Let the prime factors of n be p_i, $i = 1, 2, \ldots$. Then the sequence generated by

$$N_{k+1} = (lN_k + m) \bmod n,$$

has period n if and only if

(i) m is relatively prime to n (that is, the only common factor of m and n is 1)
(ii) $l-1$ is a multiple of p_i for all i
(iii) $l-1$ is a multiple of 4 if n is a multiple of 4.

Proof The proof of this elegant result is involved and is omitted. See Knuth (1981, p. 16).

In Example 1 above, with $l=6$, $m=7$ and $n=5$ the conditions of Theorem 5.2 were fulfilled as the reader may check. For machine computation n is often taken to be the computer's word length (2^r on an r-bit binary machine). For this reason we usually also stipulate

$$0 < l < n, \qquad 0 \leqslant m < n.$$

Special cases

For computational efficiency, m is sometimes set equal to zero, which gives a **multiplicative congruential sequence**,

$$N_{k+1} = (lN_k) \bmod n, \qquad k = 1, 2, \ldots$$

Condition (i) of Theorem 5.2 is violated but it is still possible to achieve a large period. If $n = 2^q$, where q is an integer greater than 2, the maximum period is 2^{q-2}, which is obtained if $l \equiv 3$ or $5 \pmod 8$ and N_1 is odd (Atkinson, 1980). The Control Data Corporation's random number generator (RANF) uses

$$N_{k+1} = (44485709377909 N_k) \bmod 2^{48},$$

with $u_1 = N_1/2^{48} = .170998384044023172$ (Yakowitz, 1977). This sequence has period $2^{46} = 7.0369 \times 10^{13}$. According to Knuth (1981) there are advantages in using $n = 2^\alpha - 1$. The IMSL software library employs

$$N_{k+1} = (16807 N_k) \bmod (2^{31} - 1).$$

Knuth (1981, p. 13) lists the prime factors of $2^\alpha \pm 1$ for $\alpha = 15, 16, \ldots, 64$. The number $2^{31} - 1$ is in fact a prime.

5.4 GENERATION OF RANDOM NUMBERS FROM A NORMAL DISTRIBUTION

It is necessary in many simulation studies to have 'normal random numbers' available. However, as mentioned in Section 5.2, the corollary of Theorem 5.1 is not useful for generating random numbers from a normal distribution because the inverse of the distribution function cannot be obtained explicitly. Although there are numerical routines for finding F^{-1}, it is more efficient to use what is commonly called the **polar method**. This is based on the following result (Box and Muller, 1958).

Theorem 5.3 Let U and V be independent random variables, each being uniformly distributed on $(0, 1)$. Then

$$X = (-2 \ln U)^{1/2} \cos (2\pi V),$$
$$Y = (-2 \ln U)^{1/2} \sin (2\pi V),$$

are independent standard normal random variables.

Proof Let the joint density of (X, Y) be $f_{XY}(x, y)$. If

$$U = G_1(X, Y),$$
$$V = G_2(X, Y),$$

and the joint density of (U, V) is $f_{UV}(u, v)$, then, from Chapter 1 we have

$$f_{XY}(x, y) = f_{UV}(G_1(x, y), G_2(x, y))| J(x, y)|, \tag{5.3}$$

where

$$J(x, y) = \begin{vmatrix} \dfrac{\partial G_1}{\partial x} & \dfrac{\partial G_1}{\partial y} \\ \dfrac{\partial G_2}{\partial x} & \dfrac{\partial G_2}{\partial y} \end{vmatrix},$$

is the Jacobian of the inverse transformation and $| J(x, y)|$ is its absolute value.

To apply formula (5.3) we first find U and V in terms of X and Y. Since

$$X^2 = -2 \ln U \cos^2 (2\pi V)$$
$$Y^2 = -2 \ln U \sin^2 (2\pi V),$$

we get, on adding,

$$X^2 + Y^2 = -2 \ln U.$$

On rearranging and exponentiating this becomes

$$U = G_1(X, Y) = \exp[-\tfrac{1}{2}(X^2 + Y^2)].$$

Also,

$$\frac{Y}{X} = \tan (2\pi V),$$

so

$$V = G_2(X, Y) = \frac{1}{2\pi} \arctan (Y/X).$$

In Exercise 6 it is shown that the required partial derivatives are

$$\frac{\partial G_1}{\partial x} = -x \exp[-\tfrac{1}{2}(x^2 + y^2)] \tag{5.4}$$

$$\frac{\partial G_1}{\partial y} = -y \exp[-\tfrac{1}{2}(x^2 + y^2)] \tag{5.5}$$

$$\frac{\partial G_2}{\partial x} = \frac{-y}{2\pi(x^2 + y^2)} \tag{5.6}$$

$$\frac{\partial G_2}{\partial y} = \frac{x}{2\pi(x^2 + y^2)} \tag{5.7}$$

and that the Jacobian is

$$J(x, y) = -\frac{1}{2\pi} \exp[-\tfrac{1}{2}(x^2 + y^2)].$$

Furthermore, since U, V are independent and uniformly distributed on $(0, 1)$,

$$f_{UV}(u, v) = 1, \qquad u \in (0, 1), \qquad v \in (0, 1).$$

Hence the joint density of X, Y is, on substituting in (5.3),

$$f_{XY}(x, y) = 1 \,|\, J(x, y) |$$

$$= \frac{1}{2\pi} \exp[-\tfrac{1}{2}(x^2 + y^2)], \qquad -\infty < x, y < \infty.$$

This establishes (see Section 1.6) that X and Y are independent standard normal random variables.

The polar method, based on the above theorem, is commonly used for generating normal random numbers. Another method is to utilize the central limit theorem (see Rumsey and Powner, 1975).

Various techniques must be employed for other continuous random variables. There are too many to discuss each one fully here, although some are considered in Exercises 7 and 8. The IMSL package has routines for generating random numbers from the following distributions: gamma, binomial, negative binomial, beta, Cauchy, chi-squared, geometric, exponential, hypergeometric, multinomial, log-normal, normal, Poisson, stable, triangular, uniform, Weibull, and in addition has routines for testing their properties.

5.5 STATISTICAL TESTS FOR RANDOM NUMBERS

If a simulation is to represent accurately a physical situation, it is clear that any sets of pseudo-random numbers employed must be faithful representatives of the intended random variables. Assume that random numbers have been generated which supposedly come from the distribution of a random variable X. To see if this aim has been achieved we ascertain if the numbers pass a statistical **goodness of fit test** to the required distribution. The random

numbers   must   also   be   tested   for **independence**:   the   sequence
.01, .02, .03, … would pass certain tests for a uniform distribution but would
clearly fail a test for independence.

Testing the distribution

One test of the distribution employs the χ^2-statistic (see Chapter 1). Others
include the Kolmogorov–Smirnov test which is sometimes preferable but is
not discussed here (see Chapter 3 for references). We will illustrate the use of χ^2
to test whether random numbers conform to a uniform distribution on (0, 1).
The method is easily extended to other distributions.

Suppose N random numbers have been obtained. Partition (0, 1) into K
subintervals of length $1/K$ and count the observed number of 'random'
numbers, n_k, which fall in the kth subinterval $((k-1)/K, k/K), k = 1, 2, …, K$.
On the assumption of a uniform distribution, the expected number n_k^* in each
subinterval is simply N times the length of the subinterval. That is,

$$n_k^* = N/K, \qquad k = 1, 2, …, K.$$

The χ^2-statistic is then

$$\chi^2 = \sum_{k=1}^{K} (n_k - n_k^*)^2 / n_k^*$$

$$= \frac{K}{N} \sum_{k=1}^{K} (n_k - N/K)^2.$$

There are $K - 1$ degrees of freedom because there is one linear relation,
$\sum_k n_k = N$, between the n_k. The observed value χ^2_{obs} is calculated and compared
with the critical value χ^2_{crit} at a chosen level of significance. If $\chi^2_{obs} < \chi^2_{crit}$, then it
is likely that the random numbers are from a uniform (0, 1) distribution.

Table 5.1 50 'random' numbers

.641	.773	.441	.973	.241
.814	.050	.214	.650	.614
.093	.121	.293	.921	.493
.410	.854	.010	.254	.610
.401	.013	.201	.213	.001
.294	.570	.694	.170	.094
.133	.081	.333	.881	.533
.330	.934	.930	.334	.530
.561	.853	.361	.053	.161
.974	.890	.374	.490	.774

Table 5.2 Analysis of the random numbers in the text example

k	Subinterval	n_k	$(n_k - N/K)^2$
1	(0,.1)	8	9
2	(.1,.2)	5	0
3	(.2,.3)	6	1
4	(.3,.4)	5	0
5	(.4,.5)	5	0
6	(.5,.6)	4	1
7	(.6,.7)	5	0
8	(.7,.8)	2	9
9	(.8,.9)	5	0
10	(.9,.10)	5	0
		$\Sigma = 50$	$\Sigma = 20$

Example

The first 50 numbers on (0, 1) obtained from the linear congruential sequence with $N_1 = 641, l = 123, m = 971$ and $n = 1000$ are shown in Table 5.1. The order is down columns and to the right.

To determine whether this collection is an acceptable sample from a uniform distribution on (0, 1) we divide (0, 1) into 10 subintervals as in Table 5.2 and count the number of 'random' numbers, $n_k, k = 1, \ldots, 10$, which fall in each subinterval. Here $N = 50$, $K = 10$ so $N/K = 5$ and the values of $(n_k - N/K)^2$ are easy to calculate. Table 5.2 shows the counts n_k and the values of $(n_k - N/K)^2$.

The observed value of χ^2 with 9 degrees of freedom is

$$\chi^2_9 = (10/50) \sum_{k=1}^{10} (n_k - 5)^2$$

$$= 4.$$

From the table in the Appendix we find that at the .05 level of significance the critical value of χ^2_9 is 16.919. Since $\chi^2_{obs} < \chi^2_{crit}$, we are not inclined to refute the hypothesis that the sample is from a uniform distribution on (0, 1).

5.6 TESTING FOR INDEPENDENCE

In the case of two random variables (X, Y) a random sample of size n consists of the n pairs $(X_i, Y_i), i = 1, 2, \ldots, n$. The **sample correlation coefficient** is often defined as

$$R = \frac{\sum\limits_{i=1}^{n} (X_i - \bar{X})(Y_i - \bar{Y})}{\sqrt{\left(\sum\limits_{i=1}^{n} (X_i - \bar{X})^2\right)\left(\sum\limits_{i=1}^{n} (Y_i - \bar{Y})^2\right)}}$$

where

$$\bar{X} = \frac{1}{n} \sum_{i=1}^{n} X_i,$$

$$\bar{Y} = \frac{1}{n} \sum_{i=1}^{n} Y_i,$$

are the sample means for X and Y. Values r of R close to zero indicate that X and Y are **uncorrelated**, which sometimes implies they are independent.

In the present situation we are concerned with a sequence of random variables $X_1, X_2, \ldots, X_n$. Consider pairs of consecutive variables (X_1, X_2), $(X_2, X_3), \ldots, (X_{n-1}, X_n)$. Regard the first member of each pair as one variable and the second member as another variable. The correlation coefficient of the first and second variables is called the **serial correlation coefficient at lag 1**:

$$R_1 = \frac{\sum\limits_{i=1}^{n-1} (X_i - \bar{X}_*)(X_{i+1} - \bar{X}_{**})}{\sqrt{\left(\sum\limits_{i=1}^{n-1} (X_i - \bar{X}_*)^2\right)\left(\sum\limits_{i=1}^{n-1} (X_{i+1} - \bar{X}_{**})^2\right)}} \tag{5.8}$$

where

$$\bar{X}_* = \frac{1}{n-1} \sum_{i=1}^{n-1} X_i$$

is the mean of the first $n-1$ variables and

$$\bar{X}_{**} = \frac{1}{n-1} \sum_{i=1}^{n-1} X_{i+1}$$

is the mean of the last $n-1$ variables. R_1 is also called the **autocorrelation coefficient at lag 1**. When n is large, (5.8) is usually replaced by

$$R_1 = \frac{\sum\limits_{i=1}^{n-1} (X_i - \bar{X})(X_{i+1} - \bar{X})}{\sum\limits_{i=1}^{n} (X_i - \bar{X})^2} \tag{5.9}$$

where

$$\bar{X} = \frac{1}{n} \sum_{i=1}^{n} X_i$$

is the sample mean.

It can be shown (Chatfield, 1975) that if $X_i, \ldots, X_n$ are independent, then R_1 is, for large n, approximately normally distributed with mean zero and variance $1/n$. Thus one may test for correlation between consecutive observations, which in the present context are consecutive random numbers. Under the assumption of independence we expect $|r_1| < 1.96/\sqrt{n}$ with probability approximately .95.

It is possible, however, that **consecutive** variables are independent but, for example, the pairs $(X_1, X_3), (X_2, X_4) \cdots (X_{n-2}, X_n)$ are not independent. Thus one also computes a correlation coefficient for X_i and X_{i+2}. In general, the **serial correlation coefficient at lag k** is defined as

$$R_k = \frac{\sum\limits_{i=1}^{n-k} (X_i - \bar{X})(X_{i+k} - \bar{X})}{\sum\limits_{i=1}^{n} (X_i - \bar{X})^2} \tag{5.9'}$$

Furthermore, under the assumption of independence, R_k is, when n is large, providing $k \ll n$, approximately normal with mean zero and variance $1/n$.

It is useful to plot r_k versus k to obtain a **serial correlogram**. All values of r_k must lie in the interval $[-1, 1]$ and one may draw on the correlogram the lines $\pm 1.96/\sqrt{n}$ to see which of the r_k lie outside the 95% confidence limits dictated by the assumption of independence. A hypothetical correlogram is shown in Fig. 5.3. Assuming $n = 100$, the 95% confidence limits $\pm 1.96/\sqrt{100} = \pm .196$ are indicated as dashed lines. It can be seen that r_1 and r_2 lie within the 95% limits so X_i, X_{i+1} and X_i, X_{i+2} are unlikely to be correlated. However, r_3 is too

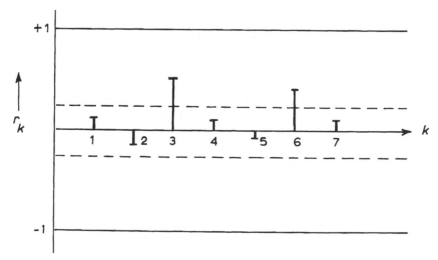

Figure 5.3 Serial correlogram with 95% confidence limits $\pm .196$.

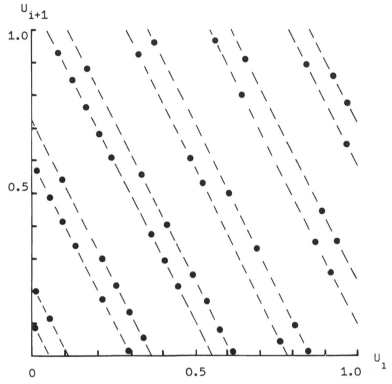

Figure 5.4 A plot u_{i+1} versus u_i for the random numbers of Table 5.1.

large and so we reject the idea that X_i and X_{i+3} are independent. Note that if $|r_k|$ is large then so too will be the absolute value of the serial correlation coefficient at lags which are multiples of k.

Before performing a serial correlation analysis it is worth while to plot X_{i+1}-values against X_i-values. This was done for the random numbers of Table 5.1 and the results are shown in Fig. 5.4. As could not be discerned by just looking at the table of random numbers, it is found that consecutive values are very closely determined by a family of relations

$$U_{i+1} = k - 2U_i,$$

where k has several different values. We may conclude immediately that the sequence of numbers is unsatisfactory. The phenomenon observed here, where the 'random' numbers lie on families of parallel straight lines, may readily occur (see Atkinson, 1980, for a complete discussion).

In addition to being useful in computer simulations, random numbers are employed in the so called Monte-Carlo method of integration (see Chapter 6 and also Yakowitz, 1977). For further reading on simulation see Morgan (1984).

REFERENCES

Atkinson, A.C. (1980). Tests of pseudo-random numbers. *Appl. Statist.*, **29**, 164–71.

Box, G.E.P. and Muller, M.E. (1958). A note on the generation of normal deviates. *Ann. Math. Statist.*, **28**, 610–11.

Chatfield, C. (1980). *The Analysis of Time Series: Theory and Practice.* Chapman and Hall, London.

Knuth, D.E. (1981). *The Art of Computer Programming*, vol. 2 Addison–Wesley. Reading, Mass.

Lehmer, D.H. (1951). Mathematical methods in large-scale computing units. In *Proc. 2nd Annual Symposium on Large-scale Digital Computing Machinery*. Harvard Univ. Press. Cambridge. Mass., pp. 141–5.

Morgan, B.J.T. (1984). *Elements of Simulation.* Chapman and Hall, London.

Rumsey, A.F. and Powner, E.T. (1975). Generation of distributions with variable envelopes. In *Digital Simulation Methods*. (ed. M.G. Hartley). Peter Peregrinus, Stevenage.

Yakowitz, S.J. (1977). *Computational Probability and Simulation.* Addison–Wesley, Reading, Mass.

EXERCISES

1. Let X have distribution function F. Prove that if F is continuous, then $F(X)$ is $U(0, 1)$.

2. Instead of using the relation $X = -(1/\lambda)\ln(1 - U)$ to generate random numbers from an exponential distribution we may use $X = -(1/\lambda)\ln U$. Why?

3. What is the maximum period attainable with a multiplicative congruential sequence with $l = 11$ and $n = 16$? What will the period be with $N_1 = 1$? $N_1 = 2$? (*Ans*: 4, 4, 2.)

4. Consider the linear congruential sequence with $l = 15$, $m = 17$, $n = 49$ and $N_1 = 15$. It is the maximum period (49) attained? What if $N_1 = 16$? (*Ans*: yes, yes.)

5. Show that in a linear congruential sequence with $n = 2^q$, m odd and $l \equiv 1$ (mod 4) the maximum period is attained.

6. Verify that the four partial derivatives of G_1 and G_2 are as given in (5.4)–(5.7). Hence verify that

$$J(x, y) = -\frac{1}{2\pi}\exp\left[-\tfrac{1}{2}(x^2 + y^2)\right].$$

7. How might random numbers for a random variable which is gamma distributed with $\rho = n$ a positive integer ≥ 2 be obtained from uniform random numbers?

8. Let X have distribution function $F(x) = x^n$, $0 \leqslant x \leqslant 1$, where n is a positive integer $\geqslant 2$. Give two methods of finding random numbers for X from uniform random numbers? (Hint: Use the result of Exercise 15, Chapter 4.)

9. Let X be a random variable with $\Pr(X = 1) = \Pr(X = -1) = \frac{1}{2}$. Consider the following sequence which is supposed to be a random sample for X:

$$\{-1, 1, -1, -1, -1, 1, -1, 1, 1, 1, -1, 1, -1, 1, 1, -1, 1, -1, 1, 1, -1\}.$$

Compute the serial (auto) correlation coefficients at lags $k = 1, 2, 3$ using (5.9'). Draw the correlogram for these values of k and mark in the 95% confidence limits $\pm 1.96/\sqrt{n}$. Is this a satisfactory random sample for X? (*Ans*: $r_1 = -0.45$, $r_2 = +0.30$, $r_3 = -0.15$. No.)

6

Convergence of sequences of random variables: the central limit theorem and the laws of large numbers

In an introductory probability course, the student will doubtless have encountered the relative frequency definition of probability and the central limit theorem. Both of these cornerstones of applied probability involve the properties of **sequences of random variables**.

In this chapter we will introduce two concepts of **convergence** of sequences of random variables. These provide the theoretical basis of the basic results mentioned as well as many others including several in statistics and the theory of random processes (see Chapters 7-10). This material is a contact point with some of the more mathematical parts of probability theory, to which references are given later in the chapter. The present discussion, however, is mostly at an elementary level.

In our study of sequences of random variables it will prove useful to employ characteristic functions, which we now define.

6.1 CHARACTERISTIC FUNCTIONS

Let X be a random variable and let t be a real number. For fixed t we may define a complex-valued random variable e^{itX}. By Euler's formula this is

$$e^{itX} = \cos(tX) + i \sin(tX).$$

The expected value of e^{itX} for various t is called the **characteristic function of X**, denoted by $\phi_X(t)$. If there is no ambiguity as to which random variable we are talking about we will drop the subscript X.

Definition Let X be a random variable. Then

$$\boxed{\phi(t) = E(e^{itX})}, \qquad -\infty < t < \infty,$$

is the characteristic function of X.

We will now investigate some of the properties of characteristic functions.

Sum of independent random variables

We will see that the characteristic function of a sum of independent random variables is the product of their characteristic functions.

Theorem 6.1 Let X_k, $k = 1, 2, \ldots, n$, be mutually independent random variables with characteristic functions

$$\phi_k(t) = E(e^{itX_k}).$$

Then the characteristic function of their sum

$$X = \sum_{k=1}^{n} X_k$$

is

$$\phi(t) = \prod_{k=1}^{n} \phi_k(t).$$

Corollary If the X_k are independent and identically distributed with characteristic function ϕ_1, then

$$\phi(t) = \phi_1^n(t).$$

Proof By definition,

$$\phi(t) = E(e^{itX})$$
$$= E(e^{it(X_1 + X_2 + \cdots + X_n)})$$
$$= E(e^{itX_1} e^{itX_2} \cdots e^{itX_n}).$$

The random variables $X_1, X_2, \ldots, X_n$ are independent and hence so are the random variables $e^{itX_1}, e^{itX_2}, \ldots, e^{itX_n}$. Since the expected value of a product of independent random variables is the product of their expected values

$$\phi(t) = E(e^{itX_1})E(e^{itX_2}) \cdots E(e^{itX_n})$$
$$= \phi_1(t)\phi_2(t) \cdots \phi_n(t)$$
$$= \prod_{k=1}^{n} \phi_k(t),$$

as required. If the random variables are, in addition to being independent, identically distributed, then $\phi_1 = \phi_2 = \cdots = \phi_n$ and the corollary follows.

Moment generating property

Repeated differentiation of a characteristic function yields the moments of the corresponding random variable as the following result indicates.

Theorem 6.2 Let X have finite nth moments m_n up to order k. Then

$$\boxed{m_n = (-1)^n i^n \phi^{(n)}(0)}, \quad n = 0, 1, \ldots, k,$$

where $\phi^{(n)}$ is the nth derivative of ϕ.

Proof Differentiating the characteristic function gives

$$\phi'(t) = \frac{d}{dt}\left(E[e^{itX}]\right) = E\left[\frac{d}{dt}(e^{itX})\right]$$

$$= E[iXe^{itX}],$$

where use has been made of the fact that the order of the operations of differentiation with respect to a parameter and expectation does not matter (see, for example, Kolmogorov, 1956). Each differentiation brings down a factor of iX, so

$$\phi^{(n)}(t) = E[(iX)^n e^{itX}], \quad n = 0, 1, 2, \ldots, k.$$

Putting $t = 0$ gives

$$\phi^{(n)}(0) = E[i^n X^n] = i^n m_n,$$

or

$$m_n = \phi^{(n)}(0)/i^n.$$

On multiplying numerator and denominator by i^n the required result is obtained.

Linear transformation

The next result shows how the characteristic function behaves under a linear transformation.

Theorem 6.3 Let X have characteristic function $\phi_X(t)$. If

$$Y = aX + b,$$

where a and b are constants, then the characteristic function of Y is

$$\boxed{\phi_Y(t) = e^{itb}\phi_X(at)}$$

Proof By definition of characteristic function,

$$\phi_Y(t) = E[e^{itY}] = E[e^{it(aX + b)}].$$

Since e^{itb} is a constant, it factors out of this expression so

$$\phi_Y(t) = e^{itb}E[e^{itaX}]$$

$$= e^{itb}\phi_X(at),$$

as required.

Uniqueness property

The reason for the term 'characteristic function' is that $\phi(t)$ **characterizes** a random variable, as the following theorem indicates.

Theorem 6.4 There is a one–one correspondence between distribution functions and characteristic functions.

Proof Proof of this theorem is beyond our scope. See for example Chung (1974).

Thus, if two random variables have characteristic functions ϕ_1 and ϕ_2 and $\phi_1(t) = \phi_2(t)$ for all t, then $F_1(x) = F_2(x)$ for all x, where F_1, F_2 are the corresponding distribution functions. As an example, if the characteristic function of a random variable can be shown to be that of a normal random variable with mean μ and variance σ^2, then the random variable is in fact normal with the stated mean and variance. This fact will be utilized in Section 6.4.

6.2 EXAMPLES

Discrete random variables

If X is discrete and takes on the values x_k with probabilities p_k, $k = 1, 2, \ldots$, then the expected value of a general function $g(X)$ of X is

$$E[g(X)] = \sum_k p_k g(x_k).$$

Hence the characteristic function of such a random variable is

$$\phi(t) = \sum_k p_k e^{itx_k}.$$

Bernoulli random variable

Let

$$\Pr\{X = 0\} = q \doteq 1 - p$$
$$\Pr\{X = 1\} = p.$$

That is, $x_1 = 0$, $p_1 = q$, $x_2 = 1$ and $p_2 = p$. Then

$$\phi(t) = p_1 e^{itx_1} + p_2 e^{itx_2}$$
$$= q + p e^{it}.$$

Binomial random variable

Let X be binomial with parameters n and p. Then X is the sum of n independent identically distributed Bernoulli variables. It follows from the

Corollary of Theorem 6.1 that the characteristic function of a binomial random variable is

$$\phi(t) = (q + pe^{it})^n. \tag{6.1}$$

Poisson random variable

Let X be Poisson with parameter λ. Then (see Exercise 1)

$$\phi(t) = \exp\left[\lambda(e^{it} - 1)\right]. \tag{6.2}$$

Continuous random variables

If X is continuous with density $f(x)$, the expected value of a general function $g(X)$ is

$$E[g(X)] = \int_{-\infty}^{\infty} g(x) f(x)\,dx.$$

Hence the characteristic function of such a random variable is

$$\phi(t) = \int_{-\infty}^{\infty} e^{itx} f(x)\,dx.$$

Exponentially distributed random variable

If X is exponentially distributed with mean $1/\lambda$, then

$$f(x) = \begin{cases} \lambda e^{-\lambda x}, & x > 0, \\ 0, & x < 0. \end{cases}$$

Hence the characteristic function is

$$\phi(t) = \lambda \int_{0}^{\infty} e^{itx} e^{-\lambda x}\,dx$$

$$= \lambda \int_{0}^{\infty} e^{-x(\lambda - it)}\,dx$$

$$= \frac{-\lambda}{\lambda - it} e^{-x(\lambda - it)}\Big|_{0}^{\infty}$$

$$= \frac{\lambda}{\lambda - it} = \frac{\lambda(\lambda + it)}{\lambda^2 + t^2}.$$

Normal random variable

Firstly we consider a standard normal random variable with mean 0 and variance 1. We will show that its characteristic function is

$$\boxed{\phi(t) = e^{-t^2/2}}.$$

Proof The standard normal density is

$$f(x) = \frac{1}{\sqrt{2\pi}} e^{-x^2/2}, \qquad -\infty < x < \infty,$$

and hence by definition

$$\phi(t) = \frac{1}{\sqrt{2\pi}} \int_{-\infty}^{\infty} e^{itx} e^{-x^2/2} \, dx.$$

Differentiating repeatedly with respect to t we find that the nth derivative of ϕ is given by

$$\phi^{(n)}(t) = \frac{i^n}{\sqrt{2\pi}} \int_{-\infty}^{\infty} x^n e^{itx} e^{-x^2/2} \, dx, \qquad n = 0, 1, 2, \ldots$$

At $t = 0$ we have

$$\phi^{(n)}(0) = \frac{i^n}{\sqrt{2\pi}} \int_{-\infty}^{\infty} x^n e^{-x^2/2} \, dx. \tag{6.3}$$

We first note that when n is odd the integrand in (6.3) is an odd function of x. Hence

$$\phi^{(n)}(0) = 0, \qquad n = 1, 3, 5, \ldots$$

When n is even put $n = 2m$, $m = 0, 1, 2, \ldots$. Then $i^n = i^{2m} = (-1)^m$ and

$$\phi^{(2m)}(0) = \frac{(-1)^m}{\sqrt{2\pi}} \int_{-\infty}^{\infty} x^{2m} e^{-x^2/2} \, dx, \qquad m = 0, 1, 2, \ldots$$

Integrating by parts we find (see Exercise 5)

$$\phi^{(2m+2)}(0) = -(2m+1)\phi^{(2m)}(0), \qquad m = 0, 1, 2, \ldots \tag{6.4}$$

Since $\phi^{(0)}(0) = 1$, being the integral of a normal density, we find

$$\phi^{(2)}(0) = -\phi^{(0)}(0) = -1,$$
$$\phi^{(4)}(0) = -3\phi^{(2)}(0) = (-1)^2 3 \cdot 1$$
$$\phi^{(6)}(0) = -5\phi^{(4)}(0) = (-1)^3 5 \cdot 3 \cdot 1$$
$$\vdots$$
$$\phi^{(2k)}(0) = (-1)^k (2k-1)(2k-3) \cdots 3 \cdot 1$$

Using the MacLaurin expansion formula and the fact that the odd-order derivatives of ϕ are zero,

$$\phi(t) = \sum_{k=0}^{\infty} \frac{\phi^{(k)}(0)t^k}{k!}$$

$$= \sum_{k=0}^{\infty} \frac{\phi^{(2k)}(0)t^{2k}}{(2k)!}$$

$$= \sum_{k=0}^{\infty} \frac{(-1)^k(2k-1)(2k-3)\cdots 3\cdot 1\cdot t^{2k}}{(2k)(2k-1)(2k-2)(2k-3)\cdots 4\cdot 3\cdot 2\cdot 1}$$

$$= \sum_{k=0}^{\infty} \frac{(-1)^k t^{2k}}{(2k)(2k-2)(2k-4)\cdots 4\cdot 2}$$

$$= \sum_{k=0}^{\infty} (-1)^k t^{2k}/2^k k!$$

This is the MacLaurin series for $e^{-t^2/2}$ so the proof is complete.

We can now prove the following.

Theorem 6.5 If X is normal with mean μ and variance σ^2, its characteristic function is

$$\phi(t) = e^{i\mu t - \sigma^2 t^2/2}$$

Proof This is Exercise 3.

6.3 CONVERGENCE IN DISTRIBUTION

When we say that a sequence of real numbers $\{x_n, n = 1, 2, \ldots\}$ converges to the limit x, we mean that for every $\varepsilon > 0$ there is an n_ε, usually depending on ε, such that

$$|x_n - x| < \varepsilon \qquad \text{for all} \qquad n > n_\varepsilon. \tag{6.5}$$

If this is true, we write

$$\lim_{n \to \infty} x_n = x.$$

Implicit in this definition is the use of the concept of **distance** between two numbers as measured in (6.5) by $|x_n - x|$: the distance between x_n and x becomes arbitrarily small for large enough n.

One way to characterize the distance between two random variables is by the 'distance' between their distribution functions.

Definition Let $\{X_n, n = 1, 2, \ldots\}$ be a sequence of random variables with distribution functions $\{F_n, n = 1, 2, \ldots\}$ and let X be a random variable with distribution function F. If

$$\lim_{n \to \infty} F_n(x) = F(x)$$

for all points x at which F is continuous, we say the sequence $\{X_n\}$ converges in distribution to X. We write

$$X_n \xrightarrow{d} X.$$

Example 1

Let $\{X_n, n = 1, 2, \ldots\}$ be a sequence of random variables which are exponentially distributed with means $(1 + 1/n)^{-1}$. Then

$$F_n(x) = 1 - e^{-(1 + 1/n)x}, \qquad x \geqslant 0. \tag{6.6}$$

At each $x \geqslant 0$ we clearly have

$$\lim_{n \to \infty} F_n(x) = 1 - e^{-x} \doteq F(x).$$

The function $F(x)$ is the distribution function of a random variable X which is exponentially distributed with mean 1. Also, $F(x)$ is continuous for all $x \in R$. Hence the sequence X_n converges in distribution to X. Some of the members of the corresponding sequence of distribution functions are shown in Fig. 6.1.

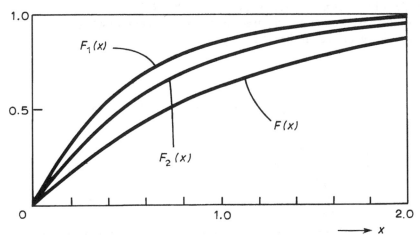

Figure 6.1 Showing how the sequence of distribution functions defined by (6.6) approaches $F(x) = 1 - e^{-x}$ as $n \to \infty$.

Example 2

This example illustrates the relevance of the phrase 'for all points x at which F is continuous' in the definition of convergence in distribution (Ash, 1970). Let X_n be uniformly distributed on $(0, 1/n)$ for $n = 1, 2, \ldots$. Let X be a random variable such that $\Pr\{X = 0\} = 1$; that is, X is the constant zero. We have

$$F_n(x) = \begin{cases} 0, & x \leqslant 0 \\ nx, & 0 < x < 1/n, \\ 1, & x \geqslant 1/n \end{cases}$$

and

$$F(x) = \begin{cases} 0, & x < 0, \\ 1, & x \geqslant 0. \end{cases}$$

Then for each $x > 0$,

$$\lim_{n \to \infty} F_n(x) = 1 = F(x).$$

$F(x)$ is not continuous at $x = 0$ nor does $F_n(0)$ approach $F(0)$ because $F_n(0) = 0$ for all n whereas $F(0) = 1$. Nevertheless $X_n \overset{d}{\to} X$ because there is convergence at all points where F is continuous. Some of the distribution functions for this example are sketched in Fig. 6.2.

There are several ways of defining convergence of a sequence of random variables. Only one other one of these, **convergence in probability**, will be discussed in this book (see Section 6.6). For a full discussion of the various modes of convergence, see, for example, Ash (1972).

In the next section we will need the following theorem whose proof can be found, for example, in the preceding reference.

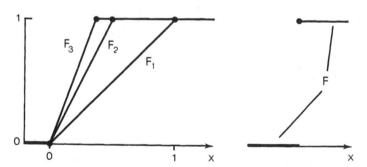

Figure 6.2 The sequence of distribution functions F_n and the limiting distribution function F in Example 2.

Theorem 6.6 The sequence of random variables $\{X_n\}$ converges in distribution to X if and only if, for all t, the sequence $\{\phi_n(t)\}$ converges to $\phi(t)$, where ϕ_n is the characteristic function of X_n and ϕ is the characteristic function of X.

Thus, to establish convergence in distribution for a sequence of random variables it is sufficient to establish the convergence of the corresponding sequence of characteristic functions. For example, let α, β be non-random constants, and suppose it is known that $X_n \overset{d}{\to} X$ and $Y_n \overset{d}{\to} Y$. If X_n, Y_n are independent for $n = 1, 2, \ldots$ and X, Y are independent, then we have

$$\phi_{\alpha X_n + \beta Y_n}(t) = \phi_{\alpha X_n}(t)\phi_{\beta Y_n}(t)$$
$$= \phi_{X_n}(\alpha t)\phi_{Y_n}(\beta t)$$
$$\to \phi_X(\alpha t)\phi_Y(\beta t)$$
$$= \phi_{\alpha X}(t)\phi_{\beta Y}(t)$$
$$= \phi_{\alpha X + \beta Y}(t).$$

Hence, by Theorem 6.6,

$$\alpha X_n + \beta Y_n \overset{d}{\to} \alpha X + \beta Y.$$

6.4 THE CENTRAL LIMIT THEOREM

There are several central limit theorems, each obtained with different assumptions on the properties of the random variables involved. The origin of the term 'central' is not clear. Some authors allude to the central role of the theorem in probability theory, while others refer to the fact that the theorem concerns a measure of central tendency, namely the mean of a sum of random variables. We will state and prove a central limit theorem which is due to Lindeberg (1922).

Theorem 6.7 (Central limit theorem) Let $\{X_k, k = 1, 2, \ldots\}$ be a sequence of independent and identically distributed random variables with means μ and variances σ^2, satisfying

$$|\mu| < \infty, \qquad 0 < \sigma^2 < \infty.$$

Define the sum

$$S_n = X_1 + X_2 + \cdots + X_n$$

and the standardized sum

$$S_n^* = \frac{S_n - E(S_n)}{\sigma(S_n)} = \frac{S_n - \mu n}{\sqrt{n}\,\sigma},$$

so that S_n^* has mean 0 and variance 1 for all n. Then

$$\boxed{S_n^* \xrightarrow[n \to \infty]{d} N(0, 1)}$$

where $N(0, 1)$ is a normal random variable with mean 0 and variance 1.

Before proving this theorem we establish the following lemma.

Lemma Let X have finite first and second moments with $E(X) = \mu$ and $\mathrm{Var}(X) = \sigma^2$. Then the characteristic function $\phi(t)$ of X has the following expansion around $t = 0$:

$$\phi(t) = 1 + i\mu t - \frac{(\mu^2 + \sigma^2)t^2}{2} + o(t^2).$$

Proof of lemma From Theorem 6.2, if the first and second moments of X exist then

$$\phi(0) = 1$$
$$\phi'(0) = i\mu$$
$$\phi''(0) = i^2 m_2 = -(\mu^2 + \sigma^2).$$

Given that $\phi'(t)$ and $\phi''(t)$ are continuous functions of t, (which may be shown true under the stated conditions) a Taylor (MacLaurin) expansion about $t = 0$ can be written using the Lagrange form of remainder (see, e.g., Rankin, 1963, or most calculus texts),

$$\phi(t) = \phi(0) + t\phi'(0) + \frac{t^2 \phi''(\theta t)}{2!}, \qquad 0 \leqslant \theta \leqslant 1.$$

The key step (see Zubrzychi, 1972) is to rewrite this by adding and subtracting the same quantity:

$$\phi(t) = \phi(0) + t\phi'(0) + \frac{t^2 \phi''(0)}{2} + \frac{t^2}{2}(\phi''(\theta t) - \phi''(0)).$$

Since ϕ'' is continuous, $\phi''(\theta t) - \phi''(0) \to 0$ as $t \to 0$ and we have the required result.

Proof of Theorem 6.7 Let the characteristic function of X_1 be $\phi(t)$. By the corollary to Theorem 6.1, the characteristic function of S_n is

$$\phi_{S_n}(t) = \phi^n(t).$$

Applying Theorem 6.3 to

$$S_n^* = \frac{S_n}{\sqrt{n\sigma}} - \frac{\mu\sqrt{n}}{\sigma}$$

we have

$$\phi_{S_n^*}(t) = \exp(-i\mu\sqrt{nt}/\sigma)\phi^n\left(\frac{t}{\sqrt{n}\sigma}\right).$$

Taking logarithms gives

$$\ln[\phi_{S_n^*}(t)] = \frac{-i\mu\sqrt{nt}}{\sigma} + n\ln\phi\left(\frac{t}{\sqrt{n}\sigma}\right).$$

By the preceding lemma this is

$$\ln[\phi_{S_n^*}(t)] = \frac{-i\mu\sqrt{nt}}{\sigma} + n\ln\left[1 + \frac{i\mu t}{\sqrt{n}\sigma} - \frac{(\mu^2+\sigma^2)t^2}{2n\sigma^2} + o\left(\frac{t^2}{n\sigma^2}\right)\right].$$

We now use the power series representation

$$\ln(1+z) = z - \frac{z^2}{2} + \frac{z^3}{3} - \cdots, \qquad |z| < 1,$$

to get

$$\ln[\phi_{S_n^*}(t)] = \frac{-i\mu\sqrt{nt}}{\sigma} + n\left[\frac{i\mu t}{\sqrt{n}\sigma} - \frac{(\mu^2+\sigma^2)t^2}{2n\sigma^2} + \frac{\mu^2 t^2}{2n\sigma^2} + o\left(\frac{t^2}{n\sigma^2}\right)\right].$$

Simplifying and rearranging,

$$\ln[\phi_{S_n^*}(t)] = \frac{-t^2}{2} + no\left(\frac{t^2}{n\sigma^2}\right)$$

But as $n \to \infty$, $1/n \to 0$ and by definition of $o(\cdot)$, $o(1/n)/(1/n) \to 0$. Hence in the limit we obtain, on exponentiating,

$$\phi_{S_n^*}(t) \xrightarrow[n\to\infty]{} e^{-t^2/2}.$$

The right-hand side is the characteristic function of a standard normal random variable. Using Theorems 6.4 and 6.6, the proof is complete.

When the X_k are independent Bernoulli random variables with

$$\Pr\{X_k = 1\} = p, \qquad 0 < p < 1,$$
$$\Pr\{X_k = 0\} = q = 1 - p,$$

the following classical result is obtained.

Theorem 6.8 (DeMoivre–Laplace form of the central limit theorem)

$$S_n^* = \frac{S_n - np}{\sqrt{npq}} \xrightarrow{d} N(0, 1).$$

Proof This result follows directly from Theorem 6.7 since it is just a special case. It can also be proved (see Exercise 9) by utilizing the known explicit form of $\phi_{S_n^*}(t)$.

The form of the central limit theorem we have proven is not the most general one. When specified supplementary conditions are satisfied, the assumptions:

(a) that the X_k are identically distributed;
(b) that the X_k are independent,

can be relaxed. See for example Zubrzychi (1972) or Ash (1972).

As will probably be known to the student from introductory courses, the central limit theorem enables approximations to be obtained for the distributions of sums of a sufficient number of random variables. In particular, it is often used to obtain approximations for various large-sample statistics. Some of these approximations are examined in the exercises.

6.5 THE POISSON APPROXIMATION TO THE BINOMIAL DISTRIBUTION

Theorem 6.9 Let X_n, $n = 1, 2, \ldots$, be binomial random variables with parameters n and p and let $np = \lambda$, where $\lambda > 0$ is fixed. Then, as $n \to \infty$,

$$X_n \overset{d}{\to} X,$$

where X is a Poisson random variable with parameter λ.

Proof From (6.1), the characteristic function of X_n is

$$\phi_{X_n}(t) = (q + pe^{it})^n,$$

where $q = 1 - p$. Taking logarithms gives

$$\ln[\phi_{X_n}(t)] = n \ln(q + pe^{it})$$
$$= n \ln(1 + p(e^{it} - 1))$$
$$= n \ln\left[1 + \frac{\lambda}{n}(e^{it} - 1)\right].$$

When n is large enough we may again use $\ln(1 + z) = z - z^2/2 + \cdots$ to get

$$\ln[\phi_{X_n}(t)] = n\left[\frac{\lambda}{n}(e^{it} - 1) - \frac{\lambda^2}{2n^2}(e^{it} - 1)^2 + \cdots\right]$$

$$\xrightarrow[n \to \infty]{} \lambda(e^{it} - 1).$$

Hence

$$\phi_{X_n}(t) \xrightarrow[n \to \infty]{} \exp[\lambda(e^{it} - 1)] = \phi_X(t),$$

which is the characteristic function of a Poisson random variable with parameter λ (see (6.2)). Invoking Theorems 6.4 and 6.6, the proof is complete.

6.6 CONVERGENCE IN PROBABILITY

In Section 6.3 we saw that one way to characterize the distance separating two random variables is by the differences between their distribution functions. Another way is as follows. Let X and Y be two random variables. The absolute value of their difference is a third random variable

$$Z = |X - Y|.$$

We may say that X and Y are close if Z is likely to be small; that is, the probability is large that their difference is small.

It is important to distinguish this from a statement about the closeness of the distribution functions of X and Y. In fact if X and Y have the same distribution functions, $|X - Y|$ may have a large fraction of its probability mass away from zero. An example was encountered in Section 2.2 where X and Y were independent and both uniformly distributed on $(0, 1)$. In that instance we found the probability density function of $Z = |X - Y|$ to be

$$f_Z(z) = 2(1 - z), \qquad 0 < z < 1.$$

To determine how close $Z = |X - Y|$ is to zero we consider its distribution function:

$$F(\varepsilon) = \Pr\{|X - Y| \leqslant \varepsilon\}.$$

If Z is equal to the constant zero, then

$$F(\varepsilon) = \begin{cases} 0, & \varepsilon < 0 \\ 1, & \varepsilon \geqslant 0. \end{cases}$$

We are thus led to the following definition of convergence.

Definition The sequence of random variables $\{X_n, n = 1, 2, \ldots\}$ is said to converge in probability to the random variable X if

$$\lim_{n \to \infty} \Pr\{|X_n - X| \leqslant \varepsilon\} = 1, \tag{6.7}$$

for all $\varepsilon > 0$. We write

$$X_n \overset{\text{P}}{\to} X.$$

This mode of convergence is also called **stochastic convergence**. Note that rather than (6.7), the equivalent statement

$$\lim_{n \to \infty} \Pr\{|X_n - X| > \varepsilon\} = 0,$$

can be, and often is, employed in the definition.

Example

For $n = 1, 2, \ldots$ suppose X_n is a random variable such that

$$X_n = \begin{cases} 0, & \text{with probability } 1/n, \\ 1, & \text{with probability } 1 - 1/n. \end{cases}$$

Let

$$X = 1 \text{ with probability } 1.$$

The only possible values of $|X_n - X|$ are 0 and 1. We have

$$|X_n - X| = \begin{cases} 0, & \text{with probability } 1 - 1/n, \\ 1, & \text{with probability } 1/n. \end{cases}$$

The distribution function of $|X_n - X|$ is therefore

$$\Pr\{|X_n - X| \leqslant \varepsilon\} = \begin{cases} 0, & \varepsilon < 0, \\ 1 - 1/n, & 0 \leqslant \varepsilon < 1, \\ 1, & \varepsilon \geqslant 1. \end{cases}$$

These distribution functions are sketched for a few values of n in Fig. 6.3. We find, therefore that

$$\lim_{n \to \infty} \Pr\{|X_n - X| \leqslant \varepsilon\} = 1 \quad \text{for all} \quad \varepsilon > 0.$$

Hence $X_n \xrightarrow{P} X$.

In this example $\{X_n\}$ also converges in distribution to X because

$$F_n(x) = \begin{cases} 0, & x < 0 \\ 1/n, & 0 \leqslant x < 1 \\ 1, & x \geqslant 1, \end{cases}$$

$$F(x) = \begin{cases} 0, & x < 1 \\ 1, & x \geqslant 1, \end{cases}$$

so $F_n(x) \to F(x)$ for all x. However, a sequence of random variables may converge in distribution but not in probability, as will be seen in the exercises.

Figure 6.3 Distribution functions of $|X_n - X|$ for $n = 1, 2, 10$ in the text example.

In fact it can be shown that $X_n \overset{P}{\to} X$ implies $X_n \overset{d}{\to} X$ but $X_n \overset{d}{\to} X$ does not imply $X_n \overset{P}{\to} X$. We say that convergence in probability is a stronger mode of convergence than convergence in distribution, or alternatively, convergence in distribution is the weaker mode of convergence.

Although we will not prove it, one can see intuitively from the definition that if $X_n \overset{P}{\to} X$ and $Y_n \overset{P}{\to} Y$, then:

(a) $X_n \pm Y_n \overset{P}{\to} X \pm Y$

(b) $X_n Y_n \overset{P}{\to} XY$.

Our immediate concern with convergence in probability (also called **convergence in measure**) is its use in the weak law of large numbers (Section 6.8). Our proof of this result involves Chebyshev's inequality, which we now discuss.

6.7 CHEBYSHEV'S INEQUALITY

The variance of a random variable X is a measure of the dispersion of the values of X about the mean with due account taken of their various probabilities. The probability that X departs from $E(X)$ by at least ε should be larger, the larger the variance of X. The following inequality quantifies this.

Theorem 6.10 (Chebyshev's inequality) Let X have finite mean μ and variance σ^2. Then, for every $\varepsilon > 0$,

$$\boxed{\Pr\{|X - \mu| \geq \varepsilon\} \leq \frac{\sigma^2}{\varepsilon^2}} \tag{6.8}$$

Proof We will assume X has a density function $f(x)$, the proof in other cases being similar. By definition, the variance is

$$\sigma^2 = \int_{-\infty}^{\infty} (x - \mu)^2 f(x)\, dx$$

$$\geq \int_{\mu+\varepsilon}^{\infty} (x - \mu)^2 f(x)\, dx + \int_{-\infty}^{\mu-\varepsilon} (x - \mu)^2 f(x)\, dx.$$

But when $x \geq \mu + \varepsilon$ and when $x \leq \mu - \varepsilon$ we have

$$(x - \mu)^2 \geq \varepsilon^2$$

Therefore

$$\sigma^2 \geq \varepsilon^2 \left[\int_{\mu+\varepsilon}^{\infty} f(x)\, dx + \int_{-\infty}^{\mu-\varepsilon} f(x)\, dx \right].$$

But the sum of these two integrals is just the probability that X differs from its mean by more than or equal to ε. Hence

$$\sigma^2 \geq \varepsilon^2 \Pr\{|X - \mu| \geq \varepsilon\}$$

On dividing by ε^2 the required result is obtained.

Theorem 6.10 is usually referred to as Chebyshev's (also Anglicized as Chebishev, Tchebycheff or Tchebychev) inequality. However, it follows from more general results, including the following.

Theorem 6.11 Let X be a non-negative random variable. Then

$$\Pr\{X \geq \varepsilon\} \leq \frac{E(X^a)}{\varepsilon^a}, \qquad 0 < \varepsilon < \infty, \qquad 0 < a < \infty. \qquad (6.9)$$

Proof The proof, which is similar to that for Theorem 6.10, is Exercise 16.

Sometimes the upper bound provided by (6.9) may be quite close to the true value, whereas often it is quite far from it. The following examples illustrate this.

Example 1

Let X be a Poisson random variable with parameter 1. Then from (6.9) we derive the inequalities

$$\Pr\{X \geq \varepsilon\} \leq \frac{E(X)}{\varepsilon} = \frac{1}{\varepsilon}$$

$$\Pr\{X \geq \varepsilon\} \leq \frac{E(X^2)}{\varepsilon^2} = \frac{2}{\varepsilon^2}$$

With $\varepsilon = 3$, these become

$$\Pr\{X \geq 3\} \leq \tfrac{1}{3} = .333..$$
$$\Pr\{X \geq 3\} \leq \tfrac{2}{9} = .222..$$

The exact value is $\Pr\{X \geq 3\} = .197$.

Example 2

Let X be $N(0, 1)$. Then, as seen in Exercise 6,

$$E(X^{2k}) = (2k - 1)(2k - 3) \cdots 3 \cdot 1, \qquad k = 1, 2, \ldots$$

Thus

$$\Pr\{|X| > \varepsilon\} \leq \frac{(2k - 1)(2k - 3) \cdots 3 \cdot 1}{\varepsilon^{2k}}$$

When $\varepsilon = 1$, the inequality is useless, the best performance being with $k = 1$:

$$\Pr\{|X| > 1\} \leqslant 1.$$

For $\varepsilon = 3$, the best inequality is obtained with $k = 4$ or $k = 5$:

$$\Pr\{|X| > 3\} \leqslant 0.016.$$

6.8 THE WEAK LAW OF LARGE NUMBERS

There are several versions of the weak law of large numbers. We will state the one due to Markov which is the most general that can be proved using Chebyshev's inequality.

Theorem 6.12 (Weak law of large numbers) Let $\{X_k, k = 1, 2, \ldots\}$ be a sequence of random variables with finite means $\{\mu_k, k = 1, 2, \ldots\}$. Define

$$\bar{X}_n = \frac{X_1 + X_2 + \cdots + X_n}{n}$$

$$\bar{\mu}_n = \frac{\mu_1 + \mu_2 + \cdots + \mu_n}{n}.$$

Then provided

$$\lim_{n \to \infty} \frac{1}{n^2} \operatorname{Var}\left(\sum_{k=1}^{n} X_k\right) = 0, \qquad (6.10)$$

we have

$$\bar{X}_n - \bar{\mu}_n \xrightarrow{\text{P}} 0.$$

Proof First observe that

$$E(\bar{X}_n) = \bar{\mu}_n,$$

and that

$$\operatorname{Var}(\bar{X}_n) = \frac{1}{n^2} \operatorname{Var}\left(\sum_{k=1}^{n} X_k\right).$$

Then by Chebyshev's inequality (6.8),

$$\Pr\{|\bar{X}_n - \bar{\mu}_n| \geqslant \varepsilon\} \leqslant \frac{1}{n^2 \varepsilon^2} \operatorname{Var}\left(\sum_{k=1}^{n} X_k\right).$$

Under condition (6.10) the quantities on the right vanish, for any $\varepsilon > 0$, as $n \to \infty$. Hence

$$\lim_{n \to \infty} \Pr\{|\bar{X}_n - \bar{\mu}_n| \geqslant \varepsilon\} = 0, \qquad \varepsilon > 0.$$

Hence we have shown that $\bar{X}_n - \bar{\mu}_n$ converges in probability to 0.

The following corollaries are also referred to as weak laws of large numbers. That they follow from Theorem 6.12 is shown in Exercise 18.

Corollary 1 Let $\{X_k, k = 1, 2, \ldots\}$ be a sequence of independent random variables with means $\{\mu_k\}$. If

$$\lim_{n \to \infty} \frac{1}{n^2} \sum_{k=1}^{n} \mathrm{Var}(X_k) = 0,$$

then $\bar{X}_n - \bar{\mu}_n \xrightarrow{\mathrm{p}} 0$.

Note: The assumption of independence can be relaxed to pairwise independence; in fact, to pairwise uncorrelatedness. We will continue to make the independence assumption which is customary.

Corollary 2 Let $\{X_k\}$ be a sequence of independent random variables with uniformly bounded variances; that is, for some $c > 0$,

$$\mathrm{Var}(X_k) \leqslant c \quad \text{for all} \quad k.$$

Then

$$\bar{X}_n - \bar{\mu}_n \xrightarrow{\mathrm{p}} 0.$$

Corollary 3 Let $\{X_k, k = 1, 2, \ldots\}$ be a sequence of independent and identically distributed r.v.s with finite means μ and finite variances σ^2. Then

$$\bar{X}_n \xrightarrow{\mathrm{p}} \mu.$$

Applications

1. Convergence of sample means

For a random sample $\{X_1, X_2, \ldots, X_n\}$ in which the means and variances are finite, Corollary 3 states that the sample mean, $\bar{X}_n$, converges in probability to μ as $n \to \infty$. Thus the average of a sequence of observations on a random variable becomes arbitrarily close to μ with a probability arbitrarily close to 1 if n is large enough. In fact Chebyshev's inequality yields the estimate

$$\Pr\{|\bar{X}_n - \mu| \geqslant \varepsilon\} \leqslant \frac{\sigma^2}{\varepsilon^2 n}.$$

2. Relative frequency interpretation of probability

In introductory probability courses it is suggested that we appeal to intuition to make it plausible that in a sequence of random experiments the relative frequency of an event approximates more closely to the probability of that event as the sequence becomes longer. The weak law of large numbers enables this notion to be stated formally. This was first demonstrated by J. Bernoulli in 1715, whose result may be stated as the following theorem.

Theorem 6.13 In n independent performances of a random experiment, let N_A be the number of times an event A occurs. Then as $n \to \infty$,

$$\frac{N_A}{n} \xrightarrow{P} \Pr\{A\}$$

where $\Pr\{A\}$ denotes the probability of the event A.

Proof For the kth experiment, define the indicator random variable

$$X_k = \begin{cases} 1, & \text{if } A \text{ occurs} \\ 0, & \text{otherwise.} \end{cases}$$

Then the X_k are independent and identically distributed with means

$$E(X_k) = \Pr\{A\}$$

and variances

$$\text{Var}(X_k) = \Pr\{A\}(1 - \Pr\{A\}).$$

Also,

$$\frac{N_A}{n} = \frac{X_1 + X_2 + \cdots + X_n}{n}.$$

The result follows immediately from Corollary 3.

We may say a little more than the above theorem. Since

$$\text{Var}(N_A/n) = \frac{1}{n} \text{Var}(X_1)$$

$$= \frac{\Pr\{A\}(1 - \Pr\{A\})}{n},$$

we have in fact

$$\Pr\left\{ \left| \frac{N_A}{n} - \Pr\{A\} \right| \geq \varepsilon \right\} \leq \frac{\Pr\{A\}(1 - \Pr\{A\})}{\varepsilon^2 n}$$

Without any information about $\Pr\{A\}$ we can utilize the fact that the maximum value of $\Pr\{A\}(1 - \Pr\{A\})$ is $1/4$, occurring when $\Pr\{A\} = \frac{1}{2}$. Thus we always have

$$\Pr\left\{ \left| \frac{N_A}{n} - \Pr\{A\} \right| \geq \varepsilon \right\} \leq \frac{1}{4\varepsilon^2 n}$$

Given any $\varepsilon > 0$ we may therefore find an n which makes the probability that the relative frequency of A differs from $\Pr\{A\}$ by more than ε as small as desired (see Exercise 23 for examples).

3. Estimation of probability distributions

Let X be a random variable. It is required to estimate its distribution function $F(x) = \Pr\{X \leqslant x\}$ for various x. The weak law of large numbers provides a framework for doing this. For we may define for a sequence of n independent observations on X, labelled $X_1, X_2, \ldots, X_n$, the indicator random variables

$$X_k^*(x) = \begin{cases} 1, & \text{if } X_k \leqslant x, \\ 0, & \text{otherwise.} \end{cases}$$

Then

$$E(X_k^*(x)) = F(x),$$

and each $X_k^*(x)$ has a finite variance. Hence by Corollary 3, as $n \to \infty$,

$$\frac{X_1^* + X_2^* + \cdots + X_n^*}{n} \xrightarrow{P} F(x).$$

4. Monte-Carlo method of integration

Suppose $f(x)$ is a given function of a real variable x as sketched, for example, in Fig. 6.4. It is required to find the value of the integral

$$I = \int_0^a f(x)\,dx.$$

A rectangle of height b is drawn to totally enclose $f(x)$ as shown. Now let X and Y be independent random variables uniformly distributed on $(0, a)$ and $(0, b)$ respectively, so (X, Y) is jointly uniformly distributed on the rectangle $0 < x < a, 0 < y < b$. The probability that a point $P = (X, Y)$ falls in the region

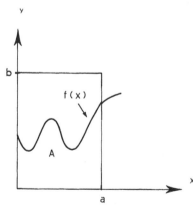

Figure 6.4 The graph of $f(x)$, a function whose integral is required.

A between $f(x)$ and the x-axis is then

$$p_A = \left(\int_0^a f(x)\,dx \right) \bigg/ ab.$$

A random sample of size n may be generated (see Chapter 5) for (X, Y). For the kth member of this sample define

$$Z_k = \begin{cases} 1, & \text{if } (X_k, Y_k) \in A, \\ 0, & \text{otherwise.} \end{cases}$$

Since $E(Z_k) = p_A$ for all k, we have by the weak law of large numbers

$$\frac{Z_1 + Z_2 + \cdots + Z_n}{n} \xrightarrow{\text{p}} p_A,$$

or, equivalently,

$$\frac{ab}{n} \sum_{k=1}^n Z_k \xrightarrow{\text{p}} \int_0^a f(x)\,dx.$$

Finally we point out that a stronger statement can be made than the convergence in probability of $\bar{X}_n$ to μ. This is the **strong law of large numbers** which is proved and discussed in more advanced treatments. (See for example Ash, 1972.) In that reference will be found the definitions and theory of stronger modes of convergence such as **mean square convergence** and **convergence almost everywhere.**

REFERENCES

Ash, R.B. (1970). *Basic Probability Theory*. Wiley, New York.

Ash, R.B. (1972). *Real Analysis and Probability*. Academic Press, New York.

Chung, K.L. (1974). *A Course in Probability Theory*. Academic Press, New York.

Hogg, R.V. and Craig, A.T. (1978). *Introduction to Mathematical Statistics*. Macmillan, New York.

Kolmogorov, A.N. (1956). *Foundations of Probability Theory*. Chelsea, New York.

Lindeberg, J. W. (1922). Eine neue Herleitung des Exponentialgesetzes in der Wahrscheinlichkeitsrechnung. *Math. Zeits.*, **15**, 211-25.

Rankin, R.A. (1963). An Introduction to Mathematical Analysis. Macmillan, New York.

Zubrzychi, S. (1972). *Lectures in Probability Theory and Mathematical Statistics*. Elsevier, New York.

EXERCISES

1. Show that the characteristic function of a Poisson random variable with parameter λ is

$$\phi(t) = \exp[\lambda(e^{it} - 1)].$$

2. Show that the characteristic function of a random variable which is gamma distributed with parameters n (a positive integer) and λ is

$$\phi(t) = \left(\frac{\lambda(\lambda + it)}{\lambda^2 + t^2} \right)^n.$$

3. Use Theorem 6.3 to obtain the characteristic function of a normal random variable with mean μ and variance σ^2 from that of a standard normal random variable.

4. Use characteristic functions to show that if X_1 and X_2 are independent Poisson random variables with parameters λ_1 and λ_2, then $X = X_1 + X_2$ is Poisson with parameter $\lambda = \lambda_1 + \lambda_2$. Generalize this to the sum of n Poisson random variables.

5. Verify the relation (6.4):

$$\phi^{(2m+2)}(0) = -(2m+1)\phi^{(2m)}(0), \qquad m = 0, 1, 2, \ldots$$

6. If X is $N(0, 1)$ show that for $k = 1, 2, 3, \ldots$,

$$E(X^{2k}) = (2k-1)(2k-3)\cdots 3\cdot 1 \doteq (2k-1)!!$$
$$E(X^{(2k-1)}) = 0.$$

7. Use characteristic functions to show that if X_1 and X_2 are independent normal random variables with means μ_1, μ_2 and variances σ_1^2, σ_2^2, then $X = X_1 + X_2$ is normal with mean $\mu = \mu_1 + \mu_2$ and variance $\sigma^2 = \sigma_1^2 + \sigma_2^2$. Generalize this to the sum of n independent normal random variables.

8. Let $\{X_n, n = 1, 2, \ldots\}$ be normal random variables with $E(X_n) = 1/n$, $Var(X_n) = 1$. Sketch the densities of X_1, X_2 and X_x. Prove that $X_n \overset{d}{\to} N(0, 1)$ as $n \to \infty$.

9. Prove the DeMoivre-Laplace form of the central limit theorem directly from the characteristic function of a binomial random variable,

$$\phi_n(t) = (q + pe^{it})^n.$$

10. Let X_1 and X_2 be independent random variables both uniformly distributed on $(-1, 1)$. Show that the density of $X = X_1 + X_2$ is

$$f(x) = (2 - |x|)/4, \quad x \in (-2, 2)$$

and that $E(X) = 0$, $Var(X) = 2/3$. Sketch the densities of: (a) X_1, (b) X, (c) a normal random variable with mean 0 and variance 2/3 (i.e., with density $\sqrt{3/4\pi} \exp(-3x^2/4)$). This illustrates that in the central limit theorem convergence to the normal distribution can be very rapid.

11. Prove that if $E(X^n) < \infty$ then $E(X^{n-1}) < \infty$, $n = 1, 2, \ldots$. In particular, if $E(X^2) < \infty$ then $E(X) < \infty$. Hence if the second moment of X is finite then so too are its mean and variance. Thus, the statement 'X has finite mean

and variance' can be replaced by 'X has a finite second moment'.

12. If for $n = 1, 2, \ldots, \Pr(X_n = 0) = 1/n = 1 - \Pr(X_n = 1)$ and $\Pr(X = 1) = 1$, prove that $X_n \xrightarrow{d} X$.

13. Let $\{X_k, k = 1, 2, \ldots\}$ be independent and identically distributed with finite means μ and variances σ^2. Prove that the sample mean $\bar{X}_n \xrightarrow{d} \mu$ as $n \to \infty$.

14. Let X_λ be a Poisson random variable with parameter λ, so that $E(X_\lambda) = \mathrm{Var}(X_\lambda) = \lambda$. Show that as $\lambda \to \infty$, $(X - \lambda)/\sqrt{\lambda} \xrightarrow{d} N(0, 1)$. This is the basis for the normal approximation to the Poisson distribution.

15. Show that if the conditions of Theorem 6.10 are satisfied, then for any $k > 0$, $\Pr\{|X - \mu| \geqslant k\sigma\} \leqslant 1/k^2$.

16. Prove Theorem 6.11.

17. If c is a real number and X is a random variable, prove that for any ε, $n > 0$, $\Pr\{|X - c| \geqslant \varepsilon\} \leqslant E(|X - c|^n)/\varepsilon^n$.

18. Prove Corollaries 1, 2 and 3 to Theorem 6.12.

19. Show that in Example 1 of Section 6.3. $X_n \xrightarrow{d} X$ but $X_n \xrightarrow{\;\;\not{p}\;\;} X$.

20. Let c be a constant. Prove that if $X_n \xrightarrow{d} c$, then $X_n \xrightarrow{p} c$. Hence give an alternative proof of Corollary 3 to Theorem 6.12 using the result of Exercise 13.

21. For $n = 1, 2, \ldots$ let $\Pr(X_n = e^n) = 1/n$ and $\Pr(X_n = 0) = 1 - 1/n$. Show that $X_n \xrightarrow{p} 0$ yet $E(X_n) \to \infty$ as $n \to \infty$.

22. Let X be uniformly distributed on $(0, 1)$ and for $n = 1, 2, \ldots$ let X_n be uniformly distributed on $(0, 1 + 1/n)$. Show that $X_n \xrightarrow{d} X$ but $X_n \xrightarrow{\;\;\not{p}\;\;} X$.

23. How many observations are required to ensure that the probability is at most 0.1 that the relative frequency of an event differs from its actual (unknown) probability by no more than 0.1? What if it is known that the probability of the event is between 0.1 and 0.2?

24. In statistics, if a sequence of random variables $\{\hat{\theta}_n, n = 1, 2, \ldots\}$ converges in probability to a parameter θ, then $\hat{\theta}_n$ is called a **consistent** estimator of θ. (See, for example, Hogg and Craig, 1978.) Show that the sample mean $\bar{X}_n$ is a consistent estimator of μ.

25. Let X be uniformly distributed on $(0, \alpha)$. Let $X_1, X_2, \ldots, X_n$ be a random sample of size n for X. Show that $Y_n = \max(X_1, X_2, \ldots, X_n)$ is a consistent estimator of α.

26. Let X have probability density

$$f(x) = \begin{cases} e^{-(x - \alpha)}, & x \geqslant \alpha \\ 0, & x < \alpha. \end{cases}$$

Let $X_1, X_2, \ldots, X_n$ be a random sample of size n for X. Show that $Y_n = \min(X_1, X_2, \ldots, X_n)$ is a consistent estimator of α.

27. Let Y_n be a sequence of random variables with $E(Y_n) \to \theta$ and $\mathrm{Var}(Y_n) \to 0$

as $n \to \infty$. Show that Y_n is a consistent estimator of θ. (*Hint*: Use Chebyshev's inequality to find an upper bound for $\Pr\{|Y_n - \theta| \geq \varepsilon\}$.)

28. Prove that if $X_n \xrightarrow{p} X$ then $X_n \xrightarrow{d} X$.

7
Simple random walks

7.1 RANDOM PROCESSES - DEFINITIONS AND CLASSIFICATIONS

Definition of random process

Physically, the term random (or stochastic) process refers to any quantity that evolves randomly in time or space. It is usually a dynamic object of some kind which varies in an unpredictable fashion. This situation is to be contrasted with that in classical mechanics whereby objects remain on fixed paths which may be predicted exactly from certain basic principles.

Mathematically, a random process is defined as a collection of random variables. The various members of the family are distinguished by different values of a parameter, α, say. The entire set of values of α, which we shall denote by A, is called an **index set** or **parameter set**. A random process is then a collection such as

$$\{X_\alpha, \alpha \in A\}$$

of random variables. The index set A may be **discrete** (finite or countably infinite) or **continuous**. The space in which the values of the random variables $\{X_\alpha\}$ lie is called the **state space**.

Usually there is some connection which unites, in some sense, the individual members of the process. Suppose a coin is tossed 3 times. Let X_k, with possible values 0 and 1, be the number of heads on the kth toss. Then the collection $\{X_1, X_2, X_3\}$ fits our definition of random process but as such is of no more interest than its individual members since each of these random variables is independent of the others. If however we introduce $Y_1 = X_1, Y_2 = X_1 + X_2$, $Y_3 = X_1 + X_2 + X_3$, so that Y_k records the number of heads up to and including the kth toss, then the collection $\{Y_k, k \in \{1, 2, 3\}\}$ is a random process which fits in with the physical concept outlined earlier. In this example the index set is $A = \{1, 2, 3\}$ (we have used k rather than α for the index) and the state space is the set $\{0, 1, 2, 3\}$.

The following two physical examples illustrate some of the possibilities for index sets and state spaces.

Examples

(i) Discrete time parameter

Let X_k be the amount of rainfall on day k with $k=0, 1, 2, \ldots$. The collection of random variables $X = \{X_k, k = 0, 1, 2, \ldots\}$ is a random process in discrete time. Since the amount of rainfall can be any non-negative number, the X_k have a continuous range. Hence X is said to have a **continuous state space**.

(ii) Continuous time parameter

Let $X(t)$ be the number of vehicles on a certain roadway at time t where $t \geqslant 0$ is measured relative to some reference time. Then the collection of random variables $X = \{X(t), t \geqslant 0\}$ is a random process in continuous time. Here the state space is discrete since the number of vehicles is a member of the discrete set $\{0, 1, 2, \ldots, N\}$ where N is the maximum number of vehicle that may be on the roadway.

Sample paths of a random process

The sequences of possible values of the family of random variables constituting a random process, taken in increasing order of time, say, are called **sample paths** (or **trajectories** or **realizations**). The various sample paths correspond to 'elementary outcomes' in the case of observations on a single random variable. It is often convenient to draw graphs of these and examples are shown in Fig. 7.1 for the cases:

(a) Discrete time - discrete state space, e.g., the number of deaths in a city due to automobile accidents on day k;
(b) Discrete time - continuous state space, e.g., the rainfall on day k;
(c) Continuous time - discrete state space, e.g., the number of vehicles on the roadway at time t;
(d) Continuous time - continuous state space, e.g., the temperature at a given location at time t.

Probabilistic description of random processes

Any random variable, X, may be characterized by its distribution function

$$F(x) = \Pr\{X \leqslant x\}, \quad -\infty \leqslant x \leqslant \infty$$

A discrete-parameter random process $\{X_k, k = 0, 1, 2, \ldots, n\}$ may be characterized by the **joint distribution function** of all the random variables involved,

$$F(x_0, x_1, \ldots, x_n) = \Pr\{X_0 \leqslant x_0, X_1 \leqslant x_1, \ldots, X_n \leqslant x_n\},$$
$$x_k \in (-\infty, \alpha), \quad k = 1, 2, \ldots, n,$$

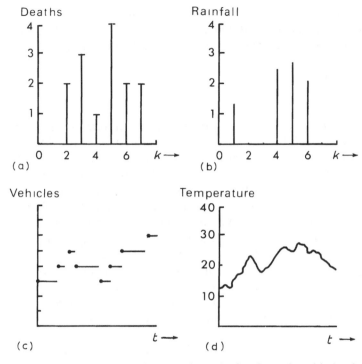

Figure 7.1 Sketches of representative sample paths for the various kinds of random processes.

and by the joint distributions of all distinct subsets of $\{X_k\}$. Similar, but more complicated descriptions apply to continuous time random processes. The probabilistic structure of some processes, however, enables them to be characterized much more simply. One important such class of processes is called **Markov processes**.

Markov processes

Definition Let $X = \{X_k, k = 0, 1, 2, \ldots\}$ be a random process with a discrete index set and a discrete state space $S = \{s_1, s_2, s_3, \ldots\}$. If

$$
\begin{aligned}
\Pr\{X_n = s_{k_n} | X_{n-1} &= s_{k_{n-1}}, X_{n-2} = s_{k_{n-2}}, \ldots, X_1 = s_{k_1}, X_0 = s_{k_0}\} \\
&= \Pr\{X_n = s_{k_n} | X_{n-1} = s_{k_{n-1}}\}
\end{aligned}
\tag{7.1}
$$

for any $n \geqslant 1$ and any collection of $s_{k_j} \in S$, $j = 0, 1, \ldots n$, then X is called a **Markov process**.

Equation (7.1) states that the values of X at all times prior to $n - 1$ have no effect whatsoever on the conditional probability distribution of X_n given X_{n-1}. Thus a Markov process has memory of its past values, but only to a limited extent.

The collection of quantities

$$\Pr\{X_n = s_{k_n} | X_{n-1} = s_{k_{n-1}}\}$$

for various n, s_{k_n} and $s_{k_{n-1}}$, is called the set of one-time-step **transition probabilities**. It will be seen later (Section 8.4) that these provide a **complete description** of the Markov process, for with them the joint distribution function of $(X_n, X_{n-1}, \ldots, X_1, X_0)$, or any subset thereof, can be found for any n. Furthermore, one only has to know the initial value of the process (in conjunction with its transition probabilities) to determine the probabilities that it will take on its various possible values at all future times. This situation may be compared with initial-value problems in differential equations, except that here probabilities are determined by the initial conditions.

All the random processes we will study in the remainder of this book are Markov processes. In the present chapter we study simple random walks which are Markov processes in discrete time and with a discrete state space. Such processes are examples of **Markov chains** which will be discussed more generally in the next chapter.

One note concerning terminology. We often talk of the **value of a process** at time t, say, which really refers to the value of a single random variable ($X(t)$), even though a process is a collection several random variables.

7.2 UNRESTRICTED SIMPLE RANDOM WALK

Suppose a particle is initially at the point $x = 0$ on the x-axis. At each subsequent time unit it moves a unit distance to the right, with probability p, or a unit distance to the left, with probability q, where $p + q = 1$.

At time unit n let the position of the particle be X_n. The above assumptions yield

$$X_0 = 0, \quad \text{with probability one,}$$

and in general,

$$X_n = X_{n-1} + Z_n, \quad n = 1, 2, \ldots,$$

where the Z_n are identically distributed with

$$\Pr\{Z_1 = +1\} = p$$
$$\Pr\{Z_1 = -1\} = q.$$

It is further assumed that the steps taken by the particle are mutually independent random variables.

Definition. The collection of random variables $X = \{X_0, X_1, X_2, \ldots\}$ is called a simple random walk in one dimension. It is 'simple' because the steps take only the values ± 1, in distinction to cases where, for example, the Z_n are continuous random variables.

The simple random walk is a random process indexed by a discrete time parameter ($n = 0, 1, 2, \ldots$) and has a discrete state space because its possible values are $\{0, \pm 1, \pm 2, \ldots\}$. Furthermore, because there are no bounds on the possible values of X, the random walk is said to be **unrestricted**.

Sample paths

Two possible beginnings of sequences of values of X are

$$\{0, +1, +2, +1, 0, -1, 0, +1, +2, +3, \ldots\}$$
$$\{0, -1, 0, -1, -2, -3, -4, -3, -4, -5, \ldots\}$$

The corresponding sample paths are sketched in Fig. 7.2.

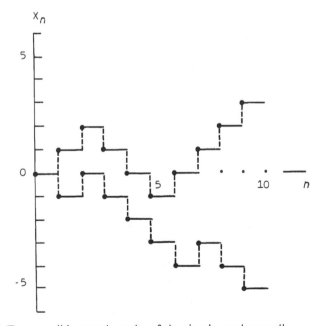

Figure 7.2 Two possible sample paths of the simple random walk.

Markov property

A simple random walk is clearly a Markov process. For example,

$$\Pr\{X_4 = 2 \mid X_3 = 3, X_2 = 2, X_1 = 1, X_0 = 0\}$$
$$= \Pr\{X_4 = 2 \mid X_3 = 3\} = \Pr\{Z_4 = +1\} = q.$$

That is, the probability is q that X_4 has the value 2 given that $X_3 = 3$, regardless of the values of the process at epochs $0, 1, 2$.

The one-time-step **transition probabilities** are

$$p_{jk} = \Pr\{X_n = k \mid X_{n-1} = j\} = \begin{cases} p, & \text{if } k = j+1 \\ q, & \text{if } k = j-1 \\ 0, & \text{otherwise} \end{cases}$$

and in this case these do not depend on n.

Mean and variance

We first observe that

$$X_1 = X_0 + Z_1$$
$$X_2 = X_1 + Z_2 = X_0 + Z_1 + Z_2$$
$$\vdots$$
$$X_n = X_0 + Z_1 + Z_2 + \cdots + Z_n.$$

Then, because the Z_n are identically distributed and independent random variables and $X_0 = 0$ with probability one,

$$E(X_n) = E\left(\sum_{k=1}^{n} Z_k\right) = nE(Z_1)$$

and

$$\operatorname{Var}(X_n) = \operatorname{Var}\left(\sum_{k=1}^{n} Z_k\right) = n\operatorname{Var}(Z_1).$$

Now,

$$E(Z_1) = 1p + (-1)q = p - q$$

and

$$E(Z_1^2) = 1p + 1q = p + q = 1.$$

Thus

$$\begin{aligned}
\operatorname{Var}(Z_1) &= E(Z_1^2) - E^2(Z_1) \\
&= 1 - (p - q)^2 \\
&= 1 - (p^2 + q^2 - 2pq) \\
&= 1 - (p^2 + q^2 + 2pq) + 4pq \\
&= 4pq.
\end{aligned}$$

since $p^2 + q^2 + 2pq = (p + q)^2 = 1$. Hence we arrive at the following expressions for the mean and variance of the process at epoch n:

$$E(X_n) = n(p - q) \qquad\qquad (7.2)$$

$$\mathrm{Var}(X_n) = 4npq \qquad\qquad (7.3)$$

We see that the mean and variance grow linearly with time.

The probability distribution of X_n

Let us derive an expression for the probability distribution of the random variable X_n, the value of the process (or x-coordinate of the particle) at time $n \geqslant 1$. That is, we seek

$$p(k, n) = \mathrm{Pr}\{X_n = k\},$$

where k is an integer.

We first note that $p(k, n) = 0$ if $n < |k|$ because the process cannot get to level k in less than $|k|$ steps. Henceforth, therefore, $n \geqslant |k|$.

Of the n steps let the number of magnitude $+1$ be N_n^+ and the number of magnitude -1 be N_n^-, where N_n^+ and N_n^- are random variables. We must have

$$X_n = N_n^+ - N_n^-$$

and

$$n = N_n^+ + N_n^-.$$

Adding these two equations to eliminate N_n^- yields

$$N_n^+ = \tfrac{1}{2}(n + X_n). \qquad\qquad (7.4)$$

Thus $X_n = k$ if and only if $N_n^+ = \tfrac{1}{2}(n + k)$. We note that N_n^+ is a binomial random variable with parameters n and p. Also, since from (7.4), $2N_n^+ = n + X_n$ is necessarily even, X_n must be even if n is even and X_n must be odd if n is odd. Thus we arrive at

$$p(k, n) = \binom{n}{(k + n)/2} p^{(k + n)/2} q^{(n - k)/2};$$

$n \geqslant |k|$, k and n either both even or both odd.

For example, the probability that the particle is at $k = -2$ after $n = 4$ steps is

$$p(-2, 4) = \binom{4}{1} pq^3 = 4pq^3. \qquad\qquad (7.5)$$

This will be verified graphically in Exercise 3.

Approximate probability distribution

If $X_0 = 0$, then

$$X_n = \sum_{k=1}^{n} Z_k,$$

where the Z_k are i.i.d. random variables with finite means and variances. Hence, by the central limit theorem (Section 6.4),

$$\frac{X_n - E(X_n)}{\sigma(X_n)} \xrightarrow{\text{d}} N(0, 1)$$

as $n \to \infty$. Since $E(X_n)$ and $\sigma(X_n)$ are known from (7.2) and (7.3), we have

$$\frac{X_n - n(p - q)}{\sqrt{4npq}} \xrightarrow{\text{d}} N(0, 1).$$

Thus for example,

$$\Pr\left\{ n(p - q) - 1.96\sqrt{4npq} < X_n < n(p - q) + 1.96\sqrt{4npq} \right\} \simeq 0.95.$$

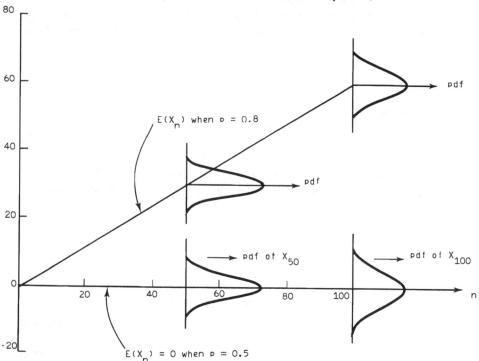

Figure 7.3 Mean of the random walk versus n for $p = 0.5$ and $p = 0.8$ and normal density approximations for the probability distributions of the process at epochs $n = 50$ and $n = 100$.

After $n = 10\,000$ steps with $p = 0.6$, $E(X_n) = 2000$ an

$$\Pr\{1808 < X_{10\,000} < 2192\} \simeq 0.95,$$

whereas when $p = 0.5$ the mean is 0 and

$$\Pr\{-196 < X_{10\,000} < 196\} \simeq 0.95.$$

Figure 7.3 shows the growth of the mean with increasing n and the approximating normal densities at $n = 50$ and $n = 100$ for various p.

7.3 RANDOM WALK WITH ABSORBING STATES

The paths of the process considered in the previous section increase or decrease at random, indefinitely. In many important applications this is not the case as particular values have special significance. This is illustrated in the following classical example.

A simple gambling game

Let two gamblers, A and B, initially have \$$a$ and \$$b$, respectively, where a and b are positive integers. Suppose that at each round of their game, player A wins \$1 from B with probability p and loses \$1 to B with probability $q = 1 - p$. The total capital of the two players at all times is

$$c = a + b.$$

Let X_n be player A's capital at round n where $n = 0, 1, 2, \ldots$ and $X_0 = a$. Let Z_n be the amount A wins on trial n. The Z_n are assumed to be independent.
    It is clear that as long as both players have money left,

$$X_n = X_{n-1} + Z_n, \qquad n = 1, 2, \ldots,$$

where the Z_n are i.i.d. as in the previous section. Thus $\{X_n, n = 0, 1, 2, \ldots\}$ is a simple random walk but there are now some restrictions or boundary conditions on the values it takes.

Absorbing states

Let us assume that A and B play until one of them has no money left; i.e., has 'gone broke'. This may occur in two ways. A's capital may reach zero or A's capital may reach c, in which case B has gone broke. The process $X = \{X_0, X_1, X_2, \ldots\}$ is thus restricted to the set of integers $\{0, 1, 2, \ldots, c\}$ and it terminates when either the value 0 or c is attained. The values 0 and c are called absorbing states, or we say there are **absorbing barriers** at 0 and c. Figure 7.4 shows plots of A's capital X_n versus trial number for two possible

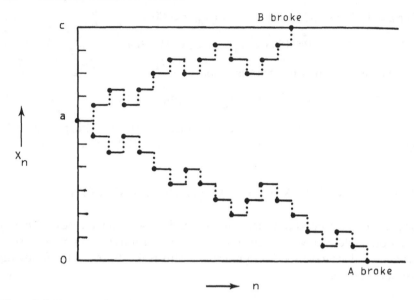

Figure 7.4 Two sample paths of a simple random walk with absorbing barriers at 0 and *c*. The upper path results in absorption at *c* (corresponding to player *A* winning all the money) and the lower one in absorption at 0 (player *A* broke).

games. One of these sample paths leads to absorption of *X* at 0 and the other to absorption at *c*.

7.4 THE PROBABILITIES OF ABSORPTION AT 0

Let P_a, $a = 0, 1, 2, \ldots, c$ denote the probabilities that player *A* goes broke when his initial capital is \$*a*. Equivalently P_a is the probability that *X* is absorbed at 0 when $X_0 = a$. The calculation of P_a is referred to as a **gambler's ruin problem**. We will obtain a difference equation for P_a.

First, however, we observe that the following boundary conditions must apply:

$$\boxed{\begin{aligned} P_0 &= 1 \\ P_c &= 0 \end{aligned}}$$

since if $a = 0$ the probability of absorption at 0 is one whereas if $a = c$, absorption at *c* has already occurred and absorption at 0 is impossible.

Now, when *a* is not equal to either 0 or *c*, all games can be divided into two mutually exclusive categories:

(i) A wins the first round;

(ii) A loses the first round.

Thus the event $\{A$ goes broke from $a\}$ is the union of two mutually exclusive events:

$\{A$ goes broke from $a\} =$
$\{A$ wins the first round and goes broke from $a + 1\}$
$\quad \cup \{A$ loses the first round and goes broke from $a - 1\}.$ (7.6)

Also, since going broke after winning the first round and winning the first round are independent,

$$\Pr\{A \text{ wins the first round and goes broke from } a + 1\}$$
$$= \Pr\{A \text{ wins the first round}\} \Pr\{A \text{ goes broke from } a + 1\}$$
$$= pP_{a+1}. \tag{7.7}$$

Similarly,

$$\Pr\{A \text{ loses the first round and goes broke from } a - 1\}$$
$$= qP_{a-1}. \tag{7.8}$$

Since the probability of the union of two mutually exclusive events is the sum of their individual probabilities, we obtain from (7.6)–(7.8), the key relation

$$\boxed{P_a = pP_{a+1} + qP_{a-1}}, \qquad a = 1, 2, \ldots, c - 1. \tag{7.9}$$

This is a difference equation for P_a which we will solve subject to the above boundary conditions.

Solution of the difference equation (7.9)

There are three main steps in solving (7.9).

(i) *The first step is to rearrange the equation*

Since $p + q = 1$, we have

$$(p + q)P_a = pP_{a+1} + qP_{a-1},$$

or

$$p(P_{a+1} - P_a) = q(P_a - P_{a-1}).$$

Dividing by p and letting

$$r = \frac{q}{p}$$

gives

$$P_{a+1} - P_a = r(P_a - P_{a-1}).$$

(ii) *The second step is to find P_1*

To do this we write out the system of equations and utilize the boundary condition $P_0 = 1$:

$$
\left.
\begin{array}{llll}
a = 1 & : & P_2 - P_1 & = r(P_1 - P_0) = r(P_1 - 1) \\
a = 2 & : & P_3 - P_2 & = r(P_2 - P_1) = r^2(P_1 - 1) \\
\vdots & & \vdots & \qquad\qquad \vdots \\
a = c - 2: & & P_{c-1} - P_{c-2} = r(P_{c-2} - P_{c-3}) = r^{c-2}(P_1 - 1) \\
a = c - 1: & & P_c - P_{c-1} = r(P_{c-1} - P_{c-2}) = r^{c-1}(P_1 - 1)
\end{array}
\right\}
\qquad (7.10)
$$

Adding all these and cancelling gives

$$
P_c - P_1 = - P_1 = (P_1 - 1)(r + r^2 + \cdots + r^{c-1}),
\qquad (7.11)
$$

where we have used the fact that $P_c = 0$.

Special case: $p = q = \frac{1}{2}$ If $p = q = \frac{1}{2}$ then $r = 1$ so $r + r^2 + \cdots + r^{c-1} = c - 1$. Hence

$$
- P_1 = (P_1 - 1)(c - 1).
$$

Solving this gives

$$
\boxed{P_1 = 1 - \frac{1}{c}}, \qquad r = 1. \qquad (7.12)
$$

General case: $p \neq q$ Equation (7.11) can be rearranged to give

$$
(P_1 - 1)(1 + r + r^2 + \cdots + r^{c-1}) + 1 = 0
$$

so

$$
P_1 = 1 - \frac{1}{1 + r + r^2 + \cdots + r^{c-1}}.
$$

For $r \neq 1$ we utilize the following formula for the sum of a finite number of terms of a geometric series:

$$
1 + r + r^2 + \cdots + r^{c-1} = \frac{1 - r^c}{1 - r}.
\qquad (7.13)
$$

Hence, after a little algebra,

$$
\boxed{P_1 = \frac{r - r^c}{1 - r^c}}, \qquad r \neq 1. \qquad (7.14)
$$

Equations (7.12) and (7.14) give the probabilities that the random walk is absorbed at zero when $X_0 = 1$, or the chances that player A goes broke when starting with one unit of capital.

(iii) *The third and final step is to solve for P_a, $a \neq 1$.*

From the system of equations (7.10) we get

$$P_2 = P_1 + r(P_1 - 1)$$
$$P_3 = P_2 + r^2(P_1 - 1) \qquad = P_1 + (P_1 - 1)(r + r^2)$$
$$\vdots \qquad \vdots \qquad \qquad \vdots$$
$$P_a = P_{a-1} + r^{a-1}(P_1 - 1) = P_1 + (P_1 - 1)(r + r^2 + \cdots + r^{a-1}).$$

Adding and subtracting one gives

$$P_a = (P_1 - 1)(1 + r + r^2 + \cdots + r^{a-1}) + 1. \qquad (7.15)$$

Special case: $p = q = \frac{1}{2}$ When $r = 1$ we have $1 + r + r^2 + \cdots + r^{a-1} = a$, so using (7.12) gives

$$\boxed{P_a = 1 - \frac{a}{c}, \qquad p = q.} \qquad (7.16)$$

General case: $p \neq q$ From (7.14) we find

$$P_1 - 1 = \frac{r - 1}{1 - r^c}.$$

Substituting this in (7.15) and utilizing (7.13) for the sum of the geometric series,

$$P_a = \left(\frac{r - 1}{1 - r^c}\right)\left(\frac{1 - r^a}{1 - r}\right) + 1,$$

which rearranges to

$$\boxed{P_a = \frac{r^a - r^c}{1 - r^c}, \qquad r \neq 1.}$$

Thus, in terms of p and q we finally obtain the following results.

Theorem 7.1 The probability that the random walk is absorbed at 0 when it starts at $X_0 = a$, (or the chances that player A goes broke from a) is

$$\boxed{P_a = \frac{(q/p)^a - (q/p)^c}{1 - (q/p)^c}, \qquad p \neq q.} \qquad (7.17)$$

Table 7.1 Values of P_a for various values of p.

a	$p = 0.25$	$p = 0.4$	$p = 0.5$
0	1	1	1
1	0.99997	0.99118	0.9
2	0.99986	0.97794	0.8
3	0.99956	0.95809	0.7
4	0.99865	0.92831	0.6
5	0.99590	0.88364	0.5
6	0.98767	0.81663	0.4
7	0.96298	0.71612	0.3
8	0.88890	0.56536	0.2
9	0.66667	0.33922	0.1
10	0	0	0

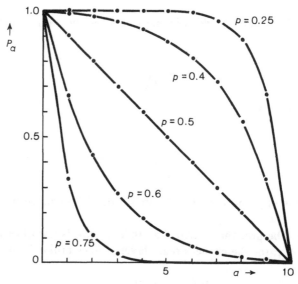

Figure 7.5 The probabilities P_a that player A goes broke. The total capital of both players is 10, a is the initial capital of A, and p = chance that A wins each round.

When $p = q = \frac{1}{2}$,

$$P_a = 1 - \frac{a}{c}.$$

Some numerical values

Table 7.1 lists values of P_a for $c = 10$, $a = 0, 1, \ldots, 10$ for the three values $p = 0.25$, $p = 0.4$ and $p = 0.5$. The values of P_a are plotted against a in Fig. 7.5. Also shown are curves for $p = 0.75$ and $p = 0.6$ which are obtained from the relation (see Exercise 8)

$$P_a(p) = 1 - P_{c-a}(1 - p).$$

In the case shown where $p = 0.25$, the chances are close to one that X will be absorbed at 0 (A will go broke) unless X_0 is 8 or more. Clearly the chances that A does not go broke are promoted by:

(i) a large p value, i.e. a high probability of winning each round;
(ii) a large value of X_0, i.e. a large share of the initial capital.

7.5 ABSORPTION AT $c > 0$

We have just considered the random walk $\{X_n, n = 0, 1, 2, \ldots\}$ where X_n was player A's fortune at epoch n. Let Y_n be player B's fortune at epoch n. Then $\{Y_n, n = 0, 1, 2, \ldots\}$ is a random walk with probability q of a step up and p of a step down at each time unit. Also, $Y_0 = c - a$ and if Y is absorbed at 0 then X is absorbed at c.

The quantity

$$Q_a = \Pr\{X \text{ is absorbed at } c \text{ when } X_0 = a\},$$

can therefore be obtained from the formulas for P_a by replacing a by $c - a$ and interchanging p and q.

Special case: $p = q = \frac{1}{2}$ In this case $P_a = 1 - a/c$ so $Q_a = 1 - (c - a)/c$. Hence

$$Q_a = \frac{a}{c}, \qquad p = q.$$

General case: $p \neq q$ From (7.17) we obtain

$$Q_a = \frac{(p/q)^{c-a} - (p/q)^c}{1 - (p/q)^c}.$$

Multiplying the numerator and denominator by $(q/p)^c$ and rearranging gives

$$Q_a = \frac{1 - (q/p)^a}{1 - (q/p)^c}, \qquad p \neq q.$$

In all cases we find

$$P_a + Q_a = 1 \tag{7.18}$$

Thus absorption at one or the other of the absorbing states is a certain event.

That the probabilities of absorption at 0 and at c add to unity is not obvious. One can imagine that a game might last forever, with A winning one round, B winning the next, A the next, and so on. Equation (7.18) tells us that the probability associated with such never-ending sample paths is zero. Hence sooner or later the random walk is absorbed, or in the gambling context, one of the players goes broke.

7.6 THE CASE $c = \infty$

If a, which is player A's initial capital, is kept finite and we let b become infinite, then player A is gambling against an opponent with infinite capital. Then, since $c = a + b$, c becomes infinite. The chances that player A goes broke are obtained by taking the limit $c \to \infty$ in expressions (7.16) and (7.17) for P_a. There are three cases to consider.

(i) $p > q$

Then player A has the advantage and since $q/p < 1$,

$$\lim_{c \to \infty} P_a = \lim_{c \to \infty} \frac{(q/p)^a - (q/p)^c}{1 - (q/p)^c} = (q/p)^a,$$

which is less than one.

(ii) $p = q$

Then the game is 'fair' and

$$\lim_{c \to \infty} P_a = \lim_{c \to \infty} 1 - \frac{a}{c} = 1.$$

(iii) $p < q$

Here player A is disadvantaged and

$$\lim_{c \to \infty} P_a = \lim_{c \to \infty} \frac{(q/p)^a - (q/p)^c}{1 - (q/p)^c} = 1$$

since $q/p > 1$.

Note that even when A and B have equal chances to win each round, player A goes broke for sure when player B has infinite initial capital. In casinos the situation is approximately that of a gambler playing someone with infinite capital, and, to make matters worse $p < q$ so the gambler goes broke with probability one if he keeps on playing. Casino owners are not usually referred to as gamblers!

7.7 HOW LONG WILL ABSORPTION TAKE?

In Section 7.5 we saw that the random walk X on a finite interval is certain to be absorbed at 0 or c. We now ask how long this will take.

Define the random variable

$$T_a = \text{time to absorption of } X \text{ when } X_0 = a, \qquad a = 0, 1, 2, \ldots, c.$$

The probability distribution of T_a can be found exactly (see for example Feller, 1968, Chapter 14) but we will find only the expected value of T_a:

$$D_a = E(T_a).$$

Clearly, if $a = 0$ or $a = c$, then absorption is immediate so we have the boundary conditions

$$D_0 = 0 \qquad (7.19)$$

$$D_c = 0 \qquad (7.20)$$

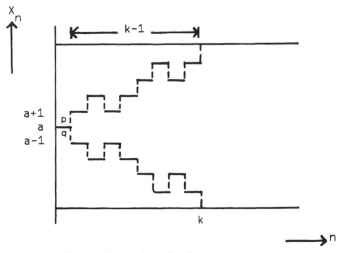

Figure 7.6 Paths leading to absorption after k steps.

We will derive a difference equation for D_a. Define

$$P(a, k) = \Pr\{T_a = k\}, \ k = 1, 2, \ldots$$

which is the probability that absorption takes k time units when the process begins at a. Considering the possible results of the first round as before (see the sketch in Fig. 7.6), we find

$$P(a, k) = pP(a + 1, k - 1) + qP(a - 1, k - 1).$$

Multiplying by k and summing over k gives

$$E(T_a) = \sum_{k=1}^{\infty} kP(a, k) = p \sum_{k=1}^{\infty} kP(a + 1, k - 1) + q \sum_{k=1}^{\infty} kP(a - 1, k - 1).$$

Putting $j = k - 1$ this may be rewritten

$$D_a = p \sum_{j=0}^{\infty} (j + 1)P(a + 1, j) + q \sum_{j=0}^{\infty} (j + 1)P(a - 1, j)$$

$$= p \sum_{j=0}^{\infty} jP(a + 1, j) + q \sum_{j=0}^{\infty} jP(a - 1, j)$$

$$+ p \sum_{j=0}^{\infty} P(a + 1, j) + q \sum_{j=0}^{\infty} P(a - 1, j).$$

But we have seen that absorption is certain, so

$$\sum_{j=0}^{\infty} P(a + 1, j) = \sum_{j=0}^{\infty} P(a - 1, j) = 1.$$

Hence

$$D_a = pD_{a+1} + qD_{a-1} + p + q$$

Table 7.2 Values of D_a from (7.22) and (7.23) with $c = 10$

a	$p = 0.25$	$p = 0.4$	$p = 0.5$
0	0	0	0
1	1.999	4.559	9
2	3.997	8.897	16
3	5.991	12.904	21
4	7.973	16.415	24
5	9.918	19.182	25
6	11.753	20.832	24
7	13.260	20.806	21
8	13.778	18.268	16
9	11.334	11.961	9
10	0	0	0

or, finally,

$$D_a = pD_{a+1} + qD_{a-1} + 1, \qquad a = 1, 2, \ldots, c - 1. \qquad (7.21)$$

This is the desired difference equation for D_a, which can be written down without the preceding steps (see Exercise 11).

The solution of (7.21) may be found in the same way that we solved the difference equation for P_a. In Exercise 12 it is found that the solution satisfying the boundary conditions (7.19), (7.20) is

$$D_a = a(c - a), \qquad p = q, \qquad (7.22)$$

$$D_a = \frac{1}{q - p}\left(a - c\left\{\frac{1 - (q/p)^a}{1 - (q/p)^c}\right\}\right), \qquad p \neq q. \qquad (7.23)$$

Numerical values

Table 7.2 lists calculated expected times to absorption for various values of a when $c = 10$ and for $p = 0.25$, $p = 0.4$ and $p = 0.5$. These values are plotted as functions of a in Fig. 7.7.

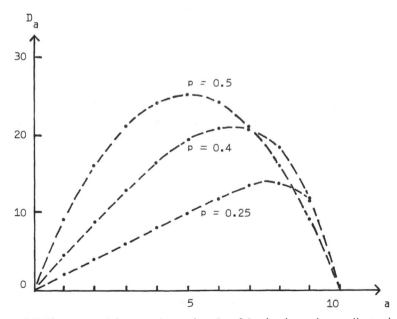

Figure 7.7 The expected times to absorption, D_a, of the simple random walk starting at a when $c = 10$ for various p.

It is seen that when $p = q$ and $c = 10$ and both players in the gambling game start with the same capital, the expected duration of the game is 25 rounds. If the total capital is $c = 1000$ and is equally shared by the two players to start with, then the average duration of their game is 250 000 rounds!

Finally we note that when $c = \infty$, the expected times to absorption are

$$
D_a = \begin{cases} \dfrac{a}{q-p}, & p < q \\ \infty, & p \geq q \end{cases}
\tag{7.24}
$$

as will be proved in Exercise 13.

7.8 SMOOTHING THE RANDOM WALK - THE WIENER PROCESS AND BROWNIAN MOTION

In Fig. 7.8a are shown portions of two possible sample paths of a simple unrestricted random walk with steps up or down of equal magnitudes. The illustrations in Fig. 7.8b–f were obtained by successive reductions of Fig. 7.8a. In (a), the 'steps' are discernible, but after several reductions the paths become smooth in appearance. In terms of the position and time scales in (a), the steps in (f) are very small and so is the time between them. The point of this is to illustrate that paths may be discontinuous but appear quite smooth when viewed from a distance.

Consider the time interval $(0, t]$. Subdivide this into subintervals of length Δt so that there are $t/\Delta t$ such subintervals. We now suppose that a particle, initially at $x = 0$, makes a step (in one space dimension) at the times $\Delta t, 2\Delta t, \dots,$ and that the size of the step is either $+\Delta x$ or $-\Delta x$, the probability being $1/2$ that the move is to the left or the right. Thus the position of the particle, $X(t)$, at time t, is a random walk which has executed $t/\Delta t$ steps. Since the position will depend on the choice of Δt and Δx, we write the position as $X(t; \Delta t, \Delta x)$.

We may write

$$
X(t; \Delta t, \Delta x) = \sum_{i=1}^{t/\Delta t} Z_i,
\tag{7.25}
$$

where the Z_i are independent and identically distributed with

$$
\Pr[Z_i = +\Delta x] = \Pr[Z_i = -\Delta x] = 1/2, \qquad i = 1, 2, \dots.
$$

For the Z_i we have,

$$
E[Z_i] = 0,
$$

and

$$
\mathrm{Var}[Z_i] = E[Z_i^2] = (\Delta x^2).
$$

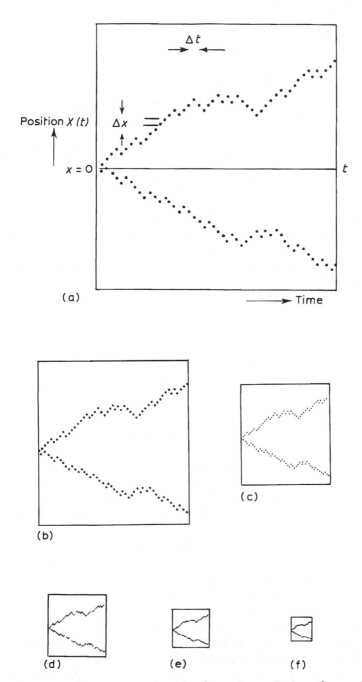

Figure 7.8 In (a) are shown two sample paths of a random walk, (b) to (f) were obtained by successive reductions of (a).

From (7.25) we get

$$E[X(t; \Delta t, \Delta x)] = 0,$$

and since the Z_i are independent,

$$\mathrm{Var}\,[X(t; \Delta t, \Delta x)] = (t/\Delta t)\,\mathrm{Var}\,[Z_i] = \frac{t(\Delta x)^2}{\Delta t}.$$

Now we let Δt and Δx get smaller so the particle moves by smaller amounts but more often. If we let Δt and Δx approach zero we won't be able to find the limiting variance as this will involve zero divided by zero, unless we prescribe a relationship between Δt and Δx.

A convenient choice is $\Delta x = \sqrt{\Delta t}$ which makes $\mathrm{Var}\,[X(t; \Delta t, \Delta x)] = t$ for all values of Δt. In the limit $\Delta t \to 0$ the random variable $X(t; \Delta t, \Delta x)$ converges in distribution to a random variable which we denote by $W(t)$. From the central limit theorem (Chapter 6) it is clear that $W(t)$ is normally distributed. Furthermore,

$$E[W(t)] = 0$$
$$\mathrm{Var}\,[W(t)] = t.$$

The collection of random variables $\{W(t), t \geqslant 0\}$, indexed by t, is a continuous process in continuous time called a **Wiener process** or **Brownian motion**, though the latter term also refers to a physical phenomenon (see below).

The Wiener process (named after Norbert Wiener, celebrated mathematician, 1894–1964) is a fascinating mathematical construction which has been much studied by mathematicians. Though it might seem just an abstraction, it has provided useful mathematical approximations to random processes in the real world. One outstanding example is Brownian motion. When a small particle is in a fluid (liquid or gas) it is buffeted around by the molecules of the fluid, usually at an astronomical rate. Each little impact moves the particle a tiny amount. You can see this if you ever watch dust or smoke particles in a stream of sunlight. This phenomenon, the erratic motion of a particle in a fluid, is called Brownian motion after the English botanist Robert Brown who observed the motion of pollen grains in a fluid under a light microscope. In 1905, Albert Einstein obtained a theory of Brownian motion using the same kind of reasoning as we did in going from random walk to Wiener process. The theory was subsequently confirmed by the experimental results of Perrin. For further reading on the Wiener process see, for example, Parzen (1962), and for more advanced aspects, Karlin and Taylor (1975) and Hida (1980).

Random walks have also been employed to represent the voltage in nerve cells (neurons). A step up in the voltage is called **excitation** and a step down is called **inhibition**. Also, there is a critical level (threshold) of excitation of which the cell emits a travelling wave of voltage called an **action potential**. The random walk model of a neuron was introduced by Gerstein and Mandelbrot (1964), who also used the Wiener process as an approximation for the voltage. Many other neural models have since been proposed and analysed (see, for example, Tuckwell, 1988).

REFERENCES

Feller, W. (1968). *An Introduction to Probability Theory and its Applications*. Wiley, New York.

Gerstein, G. L. and Mandelbrot, B. (1964). Random walk models for the spike activity of a single neuron. *Biophys. J.*, **4**, 41–68.

Hida, T. (1980). *Brownian Motion*. Springer-Verlag, New York.

Kannan, D. (1979). *An Introduction to Stochastic Processes*. North Holland, Amsterdam.

Karlin, S. and Taylor, H. (1975). *A First Course in Stochastic Processes*. Academic Press, New York.

Parzen, E. (1962). *Stochastic Processes*. Holden-Day, San Francisco.

Shiryayev, A. N. (1984). *Probability*. Springer-Verlag, New York.

Tuckwell, H. C. (1988). *Introduction to Mathematical Neurobiology*, vol. 2, *Nonlinear and Stochastic Theories*. Cambridge University Press, New York.

EXERCISES

1. Given physical examples of the four kinds of random process ((a)–(d) in Section 7.1). State in each case whether the process is a Markov process.
2. Let $X = \{X_0, X_1, X_2, \ldots\}$ be a random process in discrete time and with a discrete state space. Given that successive **increments** $X_1 - X_0$, $X_2 - X_1, \ldots$ are independent, show that X is a Markov process.
3. For a simple random walk enumerate all possible sample paths that lead to the value $X_4 = -2$ after 4 steps. Hence verify formula (7.5) for $\Pr(X_4 = -2)$.
4. Let $X_n = X_{n-1} + Z_n$, $n = 1, 2, \ldots$, describe a random walk in which the Z_n are independent normal random variables each with mean μ and variance σ^2. Find the exact probability law of X_n if $X_0 = x_0$ with probability one.
5. In certain gambling situations (e.g. horse racing, dogs) the following is an approximate description. At each trial a gambler bets m, assumed fixed. With probability q he loses all the m and with probability $p = 1 - q$ he wins back his m plus a profit on each dollar which is a random variable with mean μ and variance σ^2. Let X_n be the gambler's fortune after n bets. Deduce that $\{X_0, X_1, X_2, \ldots\}$ is a random walk with $X_0 = x_0$, the gambler's initial capital, and

$$X_n = X_{n-1} + Z_n, \qquad n = 1, 2, \ldots,$$
$$Z_n = m[I_n Y_n + (1 - I_n)],$$

where the I_n are i.i.d. indicator random variables with $\Pr(I_1 = 1) = p = 1 - \Pr(I_1 = 0)$, the Y_n are i.i.d. random variables with $E(Y_1) = \mu$, $\text{Var}(Y_1) = \sigma^2$, and it is further assumed that the I_n and Y_n are independent. Find expressions for

(i) the mean and variance of the gambler's profit on each trial;
(ii) the mean and variance of X_n;
(iii) the approximate density of X_n when n is large.

Under what conditions is there a positive expected gain on each trial?

6. Deduce that $P_a = 1 - a/c$ (equation (7.16)) when $p = q = \frac{1}{2}$ by a limiting argument from (7.17). Do this by setting $p = \frac{1}{2} + \Delta, q = \frac{1}{2} - \Delta$ and letting $\Delta \to 0$.

7. To solve the difference equation for the probability P_a that the simple random walk is absorbed at 0,

$$P_a = pP_{a+1} + qP_{a-1},$$

with boundary conditions $P_0 = 1$, $P_c = 0$, proceed as follows, the method being analogous to that of the solution of second-order differential equations with constant coefficients. Try a solution of the form $P_a = \lambda^a$. Substitute this in the difference equation to obtain a quadratic equation for λ. Let the roots of this equation be λ_1 and λ_2. General case, $p \neq q$. Write the general solution of the difference equation as

$$P_a = c_1\lambda_1{}^a + c_2\lambda_2{}^a,$$

where c_1, c_2 are determined by the boundary conditions. Special case, $p = q = \frac{1}{2}$. The general solution can be written

$$P_a = c_1\lambda^a + c_2a\lambda^a,$$

where $\lambda_1 = \lambda_2 = \lambda$ and c_1 and c_2 are determined by the boundary conditions.

8. If $P_a(p)$ denotes the probability that a simple random walk is absorbed at zero when there are absorbing states at 0 and c and the probability of a step up is p, deduce that

$$P_a(p) = 1 - P_{c-a}(1 - p).$$

9. Suppose a random walk is defined by

$$X_n = X_{n-1} + Z_n, \qquad n = 1, 2, 3, \ldots$$

where now the Z_n are i.i.d. random variables with $\Pr(Z_1 = +1) = p$, $\Pr(Z_1 = 0) = r$, $\Pr(Z_1 = -1) = q$ and $p + q + r = 1$.

(i) Sketch a few sample paths if $X_0 = 3$, $c = 8$ and there are absorbing barriers at 0 and c.
(ii) Let P_a be the probability of absorption at 0 when $X_0 = a$. Set up a

difference equation for P_a with appropriate boundary conditions at 0 and c. Hence show that the expressions for the probabilities of absorption are the same as for the simple random walk described in the text.

10. How large must a/c be when $c = 10$ to ensure no more than a 50% chance of absorption at 0 (going broke) when $p = 0.4$, $p = 0.5$ and $p = 0.6$? Repeat the calculation when $c = 20$. What conclusion do you draw?

11. Derive the difference equation (7.21) without introducing the distribution $P(a, k)$ by applying the law of total probability to expectations.

12. Solve the difference equation for the expected time to absorption of a simple random walk at 0,

$$pD_{a+1} - D_a + qD_{a-1} = -1,$$

with boundary conditions $D_0 = D_c = 0$. To do this proceed as with the solution of nonhomogeneous second-order differential equations with constant coefficients. Thus, write the general solution as

$$D_a = D_{(h)} + D_{(p)}$$

where $D_{(h)}$ is the general solution of the homogeneous equation (see Exercise 7) and $D_{(p)}$ is a particular solution of the nonhomogeneous equation. The appropriate guess for $D_{(p)}$ is ka where k is a constant found by substitution. The arbitrary constants in the general solution are obtained by applying the boundary conditions. Obtain the solution in the special case $p = q = \frac{1}{2}$ by making a limiting argument as in Exercise 6.

13. Prove that the expressions given in (7.24) apply for the expected time to absorption of the simple random walk when $c \to \infty$.

14. A random process $\{X_0, X_1, \ldots\}$ is called a **martingale** if $E(X_{n+1} \mid X_n, X_{n-1}, \ldots, X_0) = X_n$ for $n = 0, 1, 2, \ldots$. The theory of such processes is elegant and often used in probability (see, for example, Kannan, 1979, or Shiryayev, 1984). Prove that the simple random walk with $p = q$ is a martingale. An example of the usefulness of martingale results is the **optional stopping theorem**. In the context of the random walk this theorem states that the expected value of the process at the time of absorption is the same as the expected value at the beginning. If the random walk has $X_0 = a$ with probability one, then $E(X_0) = a$. Use this to deduce formula (7.16) for P_a.

15. In tennis, if the score in a game reaches deuce, a player must win two points in a row to win a game. Construe the score as a random walk on $\{0, 1, 2, 3, 4\}$. If the players have probabilities p and q of winning each point, find their probabilities of winning a game once deuce is reached. Find also the mean time to end the game.

8

Population genetics and Markov chains

8.1 GENES AND THEIR FREQUENCIES IN POPULATIONS

Chromosomes, which contain hereditary material, are located in the nuclei of the cells of living organisms. In *diploid* organisms, the chromosomes occur in pairs. For example, the nuclei of human cells contain 23 pairs of chromosomes, whereas those of dogs have 39 pairs. Sections of chromosomes that determine, by means of very complex chemical reactions, certain properties of an organism are called **genes**.

Genes may be in different forms at a given location, or **locus**, on a chromosome, and the different forms are called **alleles**. For example, in a diploid organism with two alleles labelled A_1, A_2 we may have the so called **genotypes** A_1A_1, A_1A_2, A_2A_1 or A_2A_2 as illustrated in Fig. 8.1. It is usually assumed that A_1A_2 and A_2A_1 give rise to the same properties. Individuals with both genes the same (A_1A_1 or A_2A_2) are called **homozygous**, whereas individuals with different genes (A_1A_2 or A_2A_1) are called **heterozygous**. Note that nearly all the cells of an organism have the same chromosomal structure.

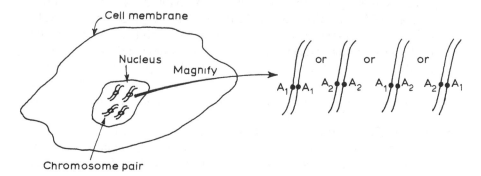

Figure 8.1 Schematic representation of a cell, its nucleus and chromosome pairs. On the right of the figure a chromosome pair is magnified to show possible genotypes at a particular locus.

In some reproductive processes (e.g. human) the chromosome pairs of the offspring contain one chromosome from each parent. Population genetics concerns itself with the numbers of genes of various types in populations, usually with a view to studying their changes from generation to generation.

Frequencies

Consider a population of N diploid individuals. At a particular locus there are a total of $2N$ genes. Let there be N_1 individuals of type A_1A_1, N_2 of type A_1A_2 or A_2A_1 and N_3 of type A_2A_2. The **genotype frequencies** are the fractions of the numbers of genotypes of each kind. We define

$$f = N_1/N$$
$$g = N_2/N$$
$$h = N_3/N,$$

and observe that

$$f + g + h = 1.$$

If we count the numbers of genes of each kind we see there are

$$p = 2N_1 + N_2$$

of type A_1, and

$$q = N_2 + 2N_3$$

of type A_2. We define the **gene (or allele) frequencies** as the fractions of the numbers of genes of each kind:

$$x = p/2N$$
$$y = q/2N.$$

We see that the relations

$$x = \frac{2N_1 + N_2}{2N} = f + g/2 \qquad (8.1)$$

$$y = \frac{N_2 + 2N_3}{2N} = g/2 + h \qquad (8.2)$$

$$x + y = 1,$$

must hold.

It should be noted that populations with different genotype frequencies may have the same gene frequencies. To illustrate consider a population of 20 individuals of which 10 are A_1A_1 and 10 are A_2A_2. Then the genotype frequencies are

$$f = 1/2, \qquad g = 0, \qquad h = 1/2,$$

while the gene frequencies are

$$x = y = 1/2.$$

If on the other hand there are $5\,A_1A_1$, $10\,A_1A_2$ and $5\,A_2A_2$, the gene frequencies are the same, but the genotype frequencies are now

$$f = 1/4, \qquad g = 1/2, \qquad h = 1/4.$$

The factors affecting the evolution of gene frequencies

We will see in a simplified picture that in 'infinite populations' of individuals which mate randomly, the gene frequencies remain constant. This is contained in the **Hardy–Weinberg principle** which we will prove in the next section. In finite populations, however, there is so called **random drift** which leads eventually to the elimination of the heterozygous genotypes. The model we use to study this phenomenon is a Markov chain and we will discuss the general introductory theory of such processes. We will then see how the properties of the Markov chain are altered when the genes themselves may change from one form to another (**mutation**). The remaining forces of evolution, namely the selective advantages of some genes over others (**selection**) and the influx or efflux of individuals (**migration**) will not be discussed here. The reader may consult Crow and Kimura (1970) or Ewens (1979) for in-depth mathematical treatments. For a fascinating account of the basic biology and biochemistry the work of Watson (1970) will be a delight to read.

8.2 THE HARDY–WEINBERG PRINCIPLE

If a population is infinite we may interpret the genotype and gene frequencies as *bona fide* (actual) probabilities. We observe such a population as it evolves in time and let the genotype frequencies at generation n be f_n, g_n and h_n and the corresponding gene frequencies be x_n and y_n, where $n = 0, 1, 2, \ldots$. We then have the following.

Theorem 8.1 (Hardy–Weinberg principle) In an infinite, randomly mating diploid population, for genes at a single locus with two alleles, the gene frequencies do not change from generation to generation. Further, no matter what the initial genotype frequencies, the genotype frequencies at the first ($n = 1$) and subsequent generations are fixed and determined only by the initial gene frequencies.

Table 8.1 Possible matings, their probabilities and the conditional probabilities of various offspring genotypes ($A_1A_2 = A_1A_2$ or A_2A_1).

Mating Male	Mating Female	$Pr(M_i)$	$Pr(A_1A_1|M_i)$	$Pr(A_1A_2|M_i)$	$Pr(A_2A_2|M_i)$
M_1	A_1A_1	A_1A_1	f_0^2	1	0	0
M_2	A_1A_1	A_1A_2	f_0g_0	$\frac{1}{2}$	$\frac{1}{2}$	0
M_3	A_1A_1	A_2A_2	f_0h_0	0	1	0
M_4	A_1A_2	A_1A_1	f_0g_0	$\frac{1}{2}$	$\frac{1}{2}$	0
M_5	A_1A_2	A_1A_2	g_0^2	$\frac{1}{4}$	$\frac{1}{2}$	$\frac{1}{4}$
M_6	A_1A_2	A_2A_2	g_0h_0	0	$\frac{1}{2}$	$\frac{1}{2}$
M_7	A_2A_2	A_1A_1	f_0h_0	0	1	0
M_8	A_2A_2	A_1A_2	g_0h_0	0	$\frac{1}{2}$	$\frac{1}{2}$
M_9	A_2A_2	A_2A_2	h_0^2	0	0	1

In symbols this becomes

(a) $x_n = x_0$, $y_n = y_0$, $n = 1, 2, 3, \ldots$
(b) $f_n = f_1$, $g_n = g_1$, $h_n = h_1$, $n = 2, 3, \ldots$
(c) f_1, g_1 and h_1 depend only upon x_0.

Proof In the population under consideration the various possible matings are shown in Table 8.1. Since both male and female parents may be any of three genotypes there are 9 possible combinations, denoted by M_i, $i = 1, 2, \ldots, 9$.

The probabilities of occurrence of the different matings are found as in the following example. The probability that an offspring of the first generation has an A_1A_1 male parent is f_0, and this is also the probability it has an A_1A_1 female parent. Hence the probability that a mating is of type M_1 is f_0^2. Similarly the remaining entries in the third column of Table 8.1 are found.

Now, given that the mating is of type M_1, the conditional probability of an A_1A_1 offspring is 1 etc. The matings M_1 - M_9 are mutually exclusive and their union is the whole sample space. Hence by the law of total probability, the probability that an offspring in the first generation is A_1A_1 is

$$f_1 = Pr\{\text{member of first generation is } A_1A_1\}$$

$$= \sum_{i=1}^{9} Pr(\text{member of first generation is } A_1A_1 | M_i) Pr(M_i)$$

$$= f_0^2 + f_0g_0 + g_0^2/4$$

$$= (f_0 + g_0/2)^2$$

Using (8.1) we get

$$f_1 = x_0^2. \tag{8.3}$$

Similarly,

$$g_1 = Pr(\text{member of first generation is } A_1A_2)$$

$$= f_0 g_0 + f_0 h_0 + g_0 h_0 + g_0^2/2$$
$$= 2(f_0 + g_0/2)(g_0/2 + h_0),$$

so that, from (8.1) and (8.2),

$$g_1 = 2x_0 y_0.$$

Also,

$$h_1 = \Pr(\text{member of first generation is } A_2 A_2)$$
$$= g_0^2/4 + g_0 h_0 + h_0^2$$
$$= (g_0/2 + h_0)^2$$
$$= y_0^2.$$

Thus we have established that the genotype frequencies in any generation are completely determined by the gene frequencies in the previous generation, regardless of the genotype frequencies in the previous generation (part (c)).

Furthermore, the frequency of A_1 in the first generation is, from (8.1),

$$x_1 = f_1 + g_1/2$$
$$= x_0^2 + x_0 y_0$$
$$= x_0(x_0 + y_0)$$
$$= x_0.$$

Hence $x_1 = x_0$ and $y_1 = y_0$. Hence the gene frequencies in any generation must be the same as those in the preceding generation (part (a)). Part (b) follows because $f_2 (= x_1^2$ from (8.3)) is determined by x_1 and $x_1 = x_0$ and so on, for $f_3, f_4, \ldots$. This completes the proof of the above form of the Hardy-Weinberg principle.

Because the gene frequencies never change and because the genotype frequencies are constant from the first generation onwards, the population is said to be in **equilibrium** or **Hardy-Weinberg equilibrium** for the gene under consideration.

An example from human genetics

The following data obtained in an actual experimental study (see Strickberger, 1968, Chapter 30) lend support to the existence of Hardy-Weinberg equilibria in nature.

When human red blood cells are injected into the bloodstreams of rabbits, an immune reaction occurs (production of antibodies) in the rabbits. However, the blood from various humans leads to different reactions and the three different human genotypes MM, MN and NN can be classified. In a group of 104 North American Indians there were found to be 61 MM, 36 MN and 7 NN individuals.

Let x be the frequency of the M allele and let $y = 1 - x$ be the frequency of the N allele. If the gene under consideration is in Hardy–Weinberg equilibrium, the genotype frequencies for MM, MN and NN should be x^2, $2xy$ and y^2 respectively. We will calculate the gene frequencies and see if the genotype frequencies are as predicted by the Hardy–Weinberg formula.

From the data we find

$$x = (122 + 36)/208 = .7596$$

and $y = .2404$. Under the hypothesis of a Hardy–Weinberg equilibrium, the expected numbers of MM, MN and NN are obtained by multiplying 104 by the genotype frequencies x^2, $2xy$ and y^2. This gives 60.01, 37.98 and 6.009, respectively.

The value of the chi-squared statistic is

$$\chi^2 = \frac{(60.01 - 61)^2}{60.01} + \frac{(37.98 - 36)^2}{37.98} + \frac{(6.009 - 7)^2}{6.006}$$

$$= 0.283.$$

There are three terms in the sum but one degree of freedom is lost because we estimated x from the data, and another degree of freedom is lost because the numbers of genotypes must add to 104. Thus there is one degree of freedom for chi-squared and from tables we find Pr $\{\chi_1^2 > 3.842\} = .05$. The observed value of chi-squared is safely less than the critical value at the .05 level of significance, lending strong support to the existence of a Hardy–Weinberg equilibrium.

8.3 RANDOM MATING IN FINITE POPULATIONS: A MARKOV CHAIN MODEL

In the previous section an infinite randomly mating diploid population was considered. The frequencies, or probabilities of occurrence of two alleles at a single locus were found to be constant.

We now wish to study the behaviour of gene frequencies in a finite population of N diploid individuals. Again we concentrate on a single locus with genotypes A_1A_1, A_1A_2 and A_2A_2. The total number of genes is fixed at $2N$ in all generations, it being assumed that the total population size is constant in time.

Notation

We introduce the following notation:

X_n = the number of A_1-genes in the nth generation, $n = 0, 1, 2, \ldots$

Thus there are $2N - X_n$ genes of type A_2 in generation n.

Random mating assumption

Randomness enters the model as follows. The $2N$ genes of any generation are chosen randomly from those in the previous generation in $2N$ Bernoulli trials in which the probability of a given gene (A_1 or A_2) is equal to its frequency in the previous generation.

Thus the number X_n of A_1-genes in generation n is a random variable and the whole sequence $X = \{X_0, X_1, X_2, \ldots\}$ is a **discrete-time random process.** Since the possible values of the X_n consist of the discrete set $\{0, 1, 2, \ldots, 2N\}$, X has a **discrete state space**. The process X is a **Markov chain**.

Transition probabilities

Suppose we are given that $X_n = j$. We ask, conditioned on this event, what is the probability that $X_{n+1} = k$. Since by the above random mating assumption, X_{n+1} is a binomial random variable with parameters $2N$ and $j/2N$, we have

$$\boxed{\Pr\{X_{n+1} = k \mid X_n = j\} = \binom{2N}{k}\left(\frac{j}{2N}\right)^k\left(1 - \frac{j}{2N}\right)^{2N-k},}$$ (8.4)

$$j, k = 0, 1, 2, \ldots, 2N.$$

This set of $(2N + 1)^2$ quantities is called the one-step transition probabilities. They can be arranged as a matrix $\mathbf{P}$ with elements

$$p_{jk} = \Pr\{X_{n+1} = k \mid X_n = j\}.$$

Before investigating the properties and behaviour of this genetical random process we give a brief introduction to the general theory of Markov chains.

8.4 GENERAL DESCRIPTION OF MARKOV CHAINS

Let $X = \{X_n, n = 0, 1, 2, \ldots\}$ be a discrete-time random process with a discrete state space $\mathscr{S}$ whose elements are $s_1, s_2, \ldots$. We have seen that X is a Markov chain if for any $n \geq 0$, the probability that X_{n+1} takes on any value $s_k \in \mathscr{S}$ is conditional only on the value of X_n (and possibly n) but does not depend on the values of $X_{n-1}, X_{n-2}, \ldots$. This leads to the introduction of the one-time-step transition probabilities

$$p_{jk}(n) = \Pr\{X_{n+1} = s_k \mid X_n = s_j\}; \qquad j, k = 1, 2, \ldots, \qquad n = 0, 1, 2, \ldots \quad (8.5)$$

We have allowed here for the possibility that the transition probabilities may depend on n. When they do not, they are called **stationary** and the process is referred to as a **temporally homogeneous** Markov chain. When they do depend on n, the term **nonhomogeneous** Markov chain is used. All the Markov chains we will consider later are temporally homogeneous.

Since X_0 is a random variable, which we refer to as the **initial value**, we introduce its probability distribution

$$p_j(0) = \Pr\{X_0 = s_j\}, \qquad j = 1, 2, \ldots \qquad (8.6)$$

We will now prove the following.

Theorem 8.2 The set of one-time-step transition probabilities (8.5) and the distribution of X_0 given by (8.6) completely determine the joint distribution of $\{X_0, X_1, \ldots, X_n\}$ for any $n \geqslant 1$.

Proof We will first prove this for $n = 1$ and $n = 2$.
$n = 1$.
We have, for any j, k, by definition of conditional probability,

$$\Pr\{X_1 = s_k | X_0 = s_j\} = \frac{\Pr\{X_0 = s_j, X_1 = s_k\}}{\Pr\{X_0 = s_j\}}.$$

On rearranging this,

$$\Pr\{X_0 = s_j, X_1 = s_k\} = \Pr\{X_0 = s_j\} \Pr\{X_1 = s_k | X_0 = s_j\}$$
$$= p_j(0) p_{jk}(1). \qquad (8.7)$$

$n = 2$.
Again by definition of conditional probability

$$\Pr\{X_2 = s_l | X_1 = s_k, X_0 = s_j\} = \frac{\Pr\{X_0 = s_j, X_1 = s_k, X_2 = s_l\}}{\Pr\{X_1 = s_k, X_0 = s_j\}},$$

so

$$\Pr\{X_0 = s_j, X_1 = s_k, X_2 = s_l\}$$
$$= \Pr\{X_1 = s_k, X_0 = s_j\} \Pr\{X_2 = s_l | X_1 = s_k, X_0 = s_j\}.$$

But $\Pr\{X_2 = s_l | X_1 = s_k, X_0 = s_j\} = \Pr\{X_2 = s_l | X_1 = s_k\}$ by the **Markov property** and so, using (8.7) as well we get

$$\Pr\{X_0 = s_j, X_1 = s_k, X_2 = s_l\} = p_j(0) p_{jk}(1) p_{kl}(2).$$

This is generalized easily to $n > 2$ (see Exercise 2).

8.5 TEMPORALLY HOMOGENEOUS MARKOV CHAINS

If a Markov chain is temporally homogeneous and there are M possible states (i.e. possible values of X), then

$$p_{jk} = \Pr\{X_{n+1} = s_k | X_n = s_j\}; \qquad j, k = 1, 2, \ldots, M, \qquad (8.8)$$

regardless of the value of n.

Definition The matrix P whose elements are given by (8.8) is called the transition matrix of the Markov chain.

Properties of P

Writing out the array **P** we have

$$\mathbf{P} = \begin{bmatrix} p_{11} & p_{12} & \cdots & p_{1M} \\ p_{21} & p_{22} & \cdots & p_{2M} \\ \cdot & \cdot & \cdots & \cdot \\ p_{M1} & p_{M2} & \cdots & p_{MM} \end{bmatrix}$$

It is seen that **P** has M rows and M columns. Every element of **P** satisfies the **non-negativity condition**

$$p_{jk} \geqslant 0. \tag{8.9}$$

Also, the sum of the elements in each row of **P** is unity. That is,

$$\sum_{k=1}^{M} p_{jk} = 1, \qquad j = 1, \ldots, M. \tag{8.10}$$

A square matrix whose elements satisfy (8.9) and (8.10) is called a **stochastic matrix**.

The probability distribution of X_n

The M quantities

$$p_j(0) = \Pr\{X_0 = s_j\}$$

can be arranged as the components of a row vector:

$$\mathbf{p}(0) = [p_1(0)\, p_2(0) \cdots p_M(0)]$$

Similarly, for X_n, $n \geqslant 1$, let

$$p_j(n) = \Pr\{X_n = s_j\},$$

and

$$\mathbf{p}(n) = [p_1(n)\, p_2(n) \cdots p_M(n)].$$

We now prove the following.

Theorem 8.3 The probability distribution of X_n, $n \geqslant 1$, is given in terms of that of X_0 by

$$\boxed{\mathbf{p}(n) = \mathbf{p}(0)\mathbf{P}^n}, \tag{8.11}$$

where **P** is the transition matrix of the Markov chain.

Proof We proceed by induction, first showing that (8.11) is true for $n = 1$.

If the value of X_0 is s_k, the value of X_1 will be s_j only if a transition is made from s_k to s_j. The events '$X_0 = s_k, k = 1, 2, \ldots, M$' are mutually exclusive and one of them must occur. Hence, by the law of total probability

$$\Pr\{X_1 = s_j\} = \sum_{k=1}^{M} \Pr\{X_0 = s_k\} \Pr\{X_1 = s_j | X_0 = s_k\},$$

or

$$p_j(1) = \sum_{k=1}^{M} p_k(0) p_{kj}, \qquad j = 1, 2, \ldots, M. \tag{8.12}$$

Recall now that if $\mathbf{A}$ is an $m \times n$ matrix with element a_{ij} in its ith row and jth column, and if $\mathbf{B}$ is an $n \times p$ matrix with general element b_{ij}, then the $m \times p$ product matrix $\mathbf{C} = \mathbf{AB}$ has general element

$$c_{ij} = \sum_{k=1}^{n} a_{ik} b_{kj}, \qquad i = 1, 2, \ldots, m; \quad j = 1, 2, \ldots, p.$$

From (8.12),

$$\mathbf{p}(1) = \mathbf{p}(0)\mathbf{P}.$$

Assume now the truth of (8.11), for some $n > 1$. Clearly

$$\Pr\{X_{n+1} = s_j\} = \sum_{k=1}^{M} \Pr\{X_n = s_k\} \Pr\{X_{n+1} = s_j | X_n = s_k\},$$

or,

$$p_j(n+1) = \sum_{k=1}^{M} p_k(n) p_{kj}.$$

In terms of vectors and matrices this becomes

$$\mathbf{p}(n+1) = \mathbf{p}(n)\mathbf{P}$$
$$= \mathbf{p}(0)\mathbf{P}^n\mathbf{P},$$

because we have assumed (8.11) is true. Since $\mathbf{P}^n\mathbf{P} = \mathbf{P}^{n+1}$, we find

$$\mathbf{p}(n+1) = \mathbf{p}(0)\mathbf{P}^{n+1}.$$

This completes the inductive proof as it follows that (8.11) is true for all $n \geqslant 1$.

The matrix $\mathbf{P}^n$ also has M rows and M columns. Its elements, denoted by $p_{jk}^{(n)}$, are called the **n-step transition probabilities** since they give the probabilities of transitions from s_j to s_k in n time steps. It is left as Exercise 3 to prove the **Chapman–Kolmogorov forward equations**

$$p_{ik}^{(m+n)} = \sum_{k=1}^{M} p_{ij}^{(m)} p_{jk}^{(n)}.$$

8.6 RANDOM GENETIC DRIFT

We now return to study the Markov chain of Section 8.3 in which X_n is the number of genes of type A_1 in a randomly mating population of size N. The state space $\mathscr{S}$ contains $2N + 1$ elements which are just the integers $0, 1, 2, \ldots, 2N$. The elements of the transition matrix are given by (8.4):

$$p_{jk} = \binom{2N}{k} \left(\frac{j}{2N} \right)^k \left(1 - \frac{j}{2N} \right)^{2N-k}; \quad j, k = 0, 1, \ldots, 2N. \quad (8.13)$$

Thus $\mathbf{P}$ has $2N + 1$ rows and $2N + 1$ columns, and in Exercise 8.4 it is shown that $\mathbf{P}$ is stochastic. For $N = 1$ the transition matrix is

$$\mathbf{P} = \tfrac{1}{4} \begin{bmatrix} 4 & 0 & 0 \\ 1 & 2 & 1 \\ 0 & 0 & 4 \end{bmatrix}.$$

When $N = 2$ we find

$$\mathbf{P} = \tfrac{1}{256} \begin{bmatrix} 256 & 0 & 0 & 0 & 0 \\ 81 & 108 & 54 & 12 & 1 \\ 16 & 64 & 96 & 64 & 16 \\ 1 & 12 & 54 & 108 & 81 \\ 0 & 0 & 0 & 0 & 256 \end{bmatrix} \quad (8.14)$$

Recall now that individuals with A_1A_1 or A_2A_2 are called homozygous whereas those with A_1A_2 are called heterozygous. We will see, first heuristically by a numerical example with $N = 2$, that a finite population of individuals which mate randomly according to our assumptions, evolves to a state in which there are no heterozygous individuals. Note that for a population of size N consisting of only homozygous individuals, the number of A_1 alleles is either 0 (corresponding to all A_2A_2) or $2N$ (all A_1A_1).

We choose a probability distribution for X_0 so that the probability that the population is homozygous is zero:

$$\mathbf{p}(0) = [0 \quad \tfrac{1}{4} \quad \tfrac{1}{2} \quad \tfrac{1}{4} \quad 0].$$

We now compute $\mathbf{p}(1) = \mathbf{p}(0)\mathbf{P}$ by matrix multiplication to find the probability distribution of X_1. This gives

$$\mathbf{p}(1) = [0.1113 \quad 0.2422 \quad 0.2930 \quad 0.2422 \quad 0.1113]$$

Similarly, the distribution of X_2 is given by $\mathbf{p}(2) = \mathbf{p}(1)\mathbf{P} = \mathbf{p}(0)\mathbf{P}^2$:

$$\mathbf{p}(2) = [0.2072 \quad 0.1868 \quad 0.2121 \quad 0.1868 \quad 0.2072]$$

The probability distributions of the number of A_1-alleles in the next four generations are found to be as follows:

$$\mathbf{p}(3) = [0.2803 \quad 0.1406 \quad 0.1583 \quad 0.1406 \quad 0.2803]$$
$$\mathbf{p}(4) = [0.3352 \quad 0.1055 \quad 0.1187 \quad 0.1055 \quad 0.3352]$$
$$\mathbf{p}(5) = [0.3764 \quad 0.0791 \quad 0.0890 \quad 0.0791 \quad 0.3764]$$
$$\mathbf{p}(6) = [0.4073 \quad 0.0593 \quad 0.0667 \quad 0.0593 \quad 0.4073].$$

Figure 8.2 shows sketches of the distributions of $X_0, X_1, \ldots, X_6$.

It can be seen that by the third generation ($n = 3$) there is more probability

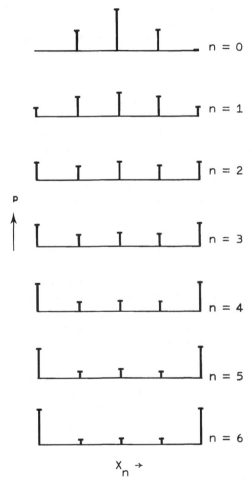

Figure 8.2 Evolution of the probability distribution of the number of A_1 genes in the text example.

mass concentrated at the homozygous states than in the heterozygous states. This contrasts with the situation in the initial population in which the probability of homozygous states was zero. By the sixth generation the probability that the population is homozygous has grown to 0.8146. Eventually, there is zero chance that the population is heterozygous, even if it started as heterozygous with probability one.

This tendency for a population to become homozygous is referred to as random genetic drift or just random drift. It was first studied theoretically by the pioneering population geneticists R.A. Fisher and Sewall Wright. This phenomenon is in direct contrast with the Hardy–Weinberg prediction of constant gene frequencies in infinite populations. It is purely due to the random sampling of gametes (egg and sperm cells) from a parent generation to form the individuals of the next generation. However, it only occurs in finite populations and the smaller the population (N), the faster is the approach to homozygosity; or, as population geneticists say, the faster do the genes become fixed in the population (all $A_1 A_1$ or all $A_2 A_2$).

Provided the assumptions which led to this theoretical prediction are fulfilled, we expect in small populations that after a few generations there is a large chance of having all homozygous individuals. In large populations the drift will proceed more slowly. The fact that nearly all Chinese have the same black hair colouring, the same brown colour eyes, etc., probably means that the population has been around a very long time and a state of homozygosity has been reached for the genes controlling these physical characteristics.

8.7 MARKOV CHAINS WITH ABSORBING STATES

Let $\{X_n, n = 0, 1, 2, \ldots\}$ be a temporally homogeneous Markov chain with state space $\mathscr{S}$ containing elements $s_1, s_2, \ldots$. Suppose it is possible to get from state s_j to state s_k in a finite time; that is, $p_{jk}^{(n)} > 0$ for some n. Then we say that state s_k is **accessible** from state s_j, or s_k can be reached from s_j. If s_j is also accessible from state s_k we say that states s_j and s_k **communicate**. A state may, of course, communicate with itself.

However, some states may act as traps so that once entered they cannot be left, as for example in the random walk of Section 7.3. If s_j is such a state, then $p_{jj} = 1$ and s_j is called **absorbing**. The jth row of the transition matrix will then consist of all zeros except for the 1 in column j. There may be just one or several absorbing states.

Absorption is certain

We make the following assumptions concerning the states of a temporally homogeneous Markov chain.

Assumptions

(i) The state space $\mathscr{S} = \{s_1, s_2, \ldots, s_M\}$ contains a finite number of elements.

(ii) The states in the set $\mathscr{A} = \{s_1, s_2, \ldots, s_A\}$, where $A \geqslant 1$, are absorbing. That is, there is at least one absorbing state.

(iii) At least one of the absorbing states is accessible from any member of the set $\mathscr{B} = \{s_{A+1}, \ldots, s_M\}$ of non-absorbing states.

We now prove the following.

Theorem 8.4 Under the above assumptions,

$$\Pr\{X_n \in \mathscr{A}\} \xrightarrow[n \to \infty]{} 1.$$

That is, absorption of X in one or other of the absorbing states is certain.

Proof If $X_0 \in \mathscr{A}$ there is nothing to prove, since X is already absorbed. Therefore, let $X_0 \in \mathscr{B}$. By assumption there is at least one state in $\mathscr{A}$ which is accessible from any state in $\mathscr{B}$. Hence, there is a state $s_k \in \mathscr{A}$ which is accessible from $s_j \in \mathscr{B}$, and so we may define $n_{jk} < \infty$ as the smallest number n such that $p_{jk}^{(n)} > 0$.

For a given state s_j let n_j be the largest of the collection of n_{jk} as k varies and let n' be the largest of the n_j as j varies. After n' time steps, no matter what the initial state of the process, there is a probability $p > 0$ that the process is in an absorbing state. Hence

$$\Pr\{X_{n'} \in \mathscr{B}\} = 1 - p$$

and $0 < (1 - p) < 1$. It follows by temporal homogeneity and the Markov property that $\Pr\{X_{2n'} \in \mathscr{B}\} \leqslant (1 - p)^2$ and, in general,

$$\Pr\{X_{kn'} \in \mathscr{B}\} \leqslant (1 - p)^k, \qquad k = 1, 2, \ldots$$

Since as $k \to \infty$, $(1 - p)^k \to 0$, we see that $\Pr\{X_n \in \mathscr{B}\} \to 0$ as $n \to \infty$. This proves that the process must eventually end up in one of the absorbing states.

Theorem 8.4 and the above proof are based on Theorem 3.1.1 of Kemeny and Snell (1960).

Example 1

For the Markov chain of Section 8.3 in which X_n is the number of genes of type A_1 in generation n, the values 0 and $2N$ are absorbing since $p_{00} = 1$ and $p_{2N,2N} = 1$. The assumptions of Theorem 8.4 are fulfilled and it follows immediately that absorption in one or the other of the absorbing states must eventually occur. That is,

$$\Pr\{X_n = 0 \cup X_n = 2N\} \xrightarrow[n \to \infty]{} 1.$$

Example 2

Consider the simple random walk of Section 7.3 where X_n is the 'position of the particle' or a 'gambler's fortune' at epoch n, with absorbing barriers at 0 and c. The elements of the transition matrix of this temporally homogeneous Markov chain are, for $j = 1, 2, \ldots, c - 1$,

$$p_{jk} = \Pr\{X_{n+1} = k \mid X_n = j\} = \begin{cases} p, & \text{if } k = j + 1 \\ q, & \text{if } k = j - 1 \\ 0, & \text{otherwise}, \end{cases}$$

whereas

$$\begin{cases} p_{00} = 1, \\ p_{0k} = 0, & k = 1, \ldots, c, \\ p_{c.k} = 0, & k = 0, \ldots, c - 1, \\ p_{cc} = 1. \end{cases}$$

Thus **P** has $c + 1$ rows and $c + 1$ columns and has the form

$$\mathbf{P} = \begin{bmatrix} 1 & 0 & 0 & 0 & \cdot & \cdot & \cdot & 0 \\ q & 0 & p & 0 & \cdot & \cdot & \cdot & 0 \\ 0 & q & 0 & p & \cdot & \cdot & \cdot & 0 \\ 0 & 0 & q & 0 & p & \cdot & \cdot & \cdot \\ \cdot & \cdot & \cdot & \cdot & \cdot & \cdot & \cdot & 0 \\ \cdot & \cdot & \cdot & \cdot & q & 0 & p \\ 0 & 0 & 0 & \cdot & \cdot & 0 & 0 & 1 \end{bmatrix}$$

It is intuitively clear that the absorbing states are accessible from any of the non-absorbing states, $1, 2, \ldots, c - 1$. By Theorem 8.4 absorption at 0 or c is certain as $n \to \infty$, a fact that we proved by a different method in Section 7.5.

8.8 ABSORPTION PROBABILITIES

Given a temporally homogeneous Markov chain which satisfies assumptions (i)–(iii) of the previous section, we have seen that the process must terminate in one of the absorbing states. If there is more than one absorbing state we may wish to know the chances of absorption in the individual absorbing states. For example, in the Markov chain model which displays random genetic drift, we would like to know the probability that the population ends up having all individuals of genotype A_1A_1 as opposed to all A_2A_2. We thus require the **absorption probabilities** for the various absorbing states. In this section we show how to calculate these probabilities as functions of the initial value of the process.

If states $s_1, \ldots, s_A$ are absorbing and there are M states altogether, the transition matrix can be put in the form

$$
\mathbf{P} = \begin{array}{c} \\ 1 \\ 2 \\ \cdot \\ \cdot \\ \cdot \\ A \\ A+1 \\ \cdot \\ \cdot \\ M \end{array}
\begin{array}{ccccccccc}
1 & 2 & \cdot & \cdot & A & A+1 & \cdot & \cdot & M \\
\left[\begin{array}{ccccc|cccc}
1 & 0 & \cdot & \cdot & 0 & 0 & \cdot & \cdot & 0 \\
0 & 1 & \cdot & \cdot & 0 & 0 & \cdot & \cdot & 0 \\
\cdot & \cdot & \cdot & & \cdot & \cdot & & & \cdot \\
\cdot & & \cdot & & \cdot & \cdot & & & \cdot \\
\cdot & & & \cdot & \cdot & \cdot & & & \cdot \\
0 & \cdot & \cdot & \cdot & 1 & 0 & \cdot & \cdot & 0 \\
\hline
p_{A+1.1} & \cdot & \cdot & \cdot & p_{A+1.A} & p_{A+1.A+1} & \cdot & \cdot & p_{A+1.M} \\
\cdot & & & & \cdot & \cdot & & & \cdot \\
\cdot & & & & \cdot & \cdot & & & \cdot \\
p_{M.1} & \cdot & \cdot & \cdot & p_{M.A} & p_{M.A+1} & \cdot & \cdot & p_{M.M}
\end{array}\right] & & & & & & & &
\end{array}
\tag{8.15}
$$

Introducing the $(M - A) \times (M - A)$ submatrix

$$
Q = \begin{bmatrix}
p_{A+1.A+1} & \cdot & \cdot & \cdot & p_{A+1.M} \\
\cdot & & & & \cdot \\
\cdot & & & & \cdot \\
\cdot & & & & \cdot \\
p_{M.A+1} & \cdot & \cdot & \cdot & p_{M.M}
\end{bmatrix}
$$

and the $(M - A) \times A$ submatrix

$$
R = \begin{bmatrix}
p_{A+1.1} & \cdot & \cdot & \cdot & p_{A+1.A} \\
\cdot & & & & \cdot \\
\cdot & & & & \cdot \\
\cdot & & & & \cdot \\
p_{M.1} & \cdot & \cdot & \cdot & p_{M.A}
\end{bmatrix}
\tag{8.16}
$$

the matrix $\mathbf{P}$ can be partitioned as

$$
\mathbf{P} = \left[\begin{array}{c|c}
\mathbf{I} & \mathbf{0} \\
\hline
\mathbf{R} & \mathbf{Q}
\end{array}\right]
$$

where $\mathbf{I}$ is an $A \times A$ identity matrix and $\mathbf{0}$ is an $A \times (M - A)$ zero matrix. The elements of $\mathbf{Q}$ are the one-step transition probabilities among the non-absorbing states, and the elements of $\mathbf{R}$ are the one-step transition probabilities from non-absorbing to absorbing states.

We now define the matrix $\mathbf{\Pi}$ whose elements are the required absorption probabilities:

$$
\pi_{jk} = \text{Pr}\{\text{process is absorbed in state}
$$
$$
s_k \in \mathscr{A} \,|\, \text{starts in } s_j \in \mathscr{B}\}
\tag{8.17}
$$

It is seen that $\mathbf{\Pi}$ has $(M - A)$ rows and A columns. We introduce the matrix

$$
\boxed{\mathbf{\Phi} = (\mathbf{I} - \mathbf{Q})^{-1}},
$$

which is called the **fundamental matrix** of the Markov chain, where here **I** is an identity matrix with the same number of rows and columns as **Q**. In terms of **Φ** and the matrix **R** defined by (8.16) we have the following result.

Theorem 8.5 The matrix whose elements are the absorption probabilities (8.17) is given by

$$\boxed{\Pi = \Phi R}$$

Proof From the state $s_j \in \mathscr{B}$ the process goes at the first time-step to state $s_i \in \mathscr{S}$ with probability p_{ji}. Allowing for these possible first transitions we have

$$\pi_{jk} = \sum_{i=1}^{M} p_{ji} \Pr\{\text{process is absorbed in state}$$
$$s_k | \text{starts in } s_i\}. \tag{8.18}$$

Allowing for the contingencies

$$\Pr\left\{\begin{array}{l}\text{process starts in state } s_i \text{ and is} \\ \text{absorbed in state } s_k\end{array}\right\} = \begin{cases} 1, & s_i = s_k, \\ 0, & s_i \in \mathscr{A}, \quad s_i \neq s_k, \\ \pi_{ik}, & s_i \in \mathscr{B}, \quad i = A+1,\ldots,M, \end{cases}$$

equation (8.18) becomes

$$\pi_{jk} = p_{jk} + \sum_{i=A+1}^{M} p_{ji}\pi_{ik}, \quad j = A+1,\ldots,M; \; k = 1,\ldots,A. \tag{8.19}$$

But $p_{jk}, j = A+1,\ldots,M; k = 1,\ldots,A$ are the elements of **R**, whereas p_{ji}, $j = A+1,\ldots,M; i = A+1,\ldots,M$ are the elements of **Q**. Hence, in matrix notation, (8.19) becomes

$$\Pi = R + Q\Pi.$$

Rearranging and being careful to preserve the order of matrix multiplication,

$$(I - Q)\Pi = R.$$

Premultiplying both sides with the inverse of $(I - Q)$ gives

$$\Pi = (I - Q)^{-1}R,$$

which proves the theorem, since $\Phi = (I - Q)^{-1}$.

Example 1

Consider the Markov chain model for the numbers of A_1-genes in a (self-fertilizing) population with $N = 1$. The possible values are 0, 1, 2. The matrix of probabilities of transitions among the non-absorbing states consists of a single entry,

$$Q = [p_{11}] = [\tfrac{1}{2}].$$

The matrix of probabilities of transitions from non-absorbing to absorbing states is

$$\mathbf{R} = [p_{10}\, p_{12}] = [\tfrac{1}{4}\,\tfrac{1}{4}].$$

Then

$$\mathbf{I} - \mathbf{Q} = [\tfrac{1}{2}]$$

and

$$\mathbf{\Phi} = (\mathbf{I} - \mathbf{Q})^{-1} = [2]. \qquad (8.20)$$

Thus, from Theorem 8.5, the absorption probabilities are

$$\mathbf{\Pi} = [\pi_{10}\, \pi_{12}] = \mathbf{\Phi R}$$
$$= [\tfrac{1}{2}\,\tfrac{1}{2}].$$

Example 2

Let a simple random walk be restricted by absorbing barriers at 0 and $c = 3$. The transition matrix is

$$\mathbf{P} = \begin{array}{c} \\ 0 \\ 1 \\ 2 \\ 3 \end{array} \begin{array}{cccc} 0 & 1 & 2 & 3 \\ \left[\begin{array}{cccc} 1 & 0 & 0 & 0 \\ q & 0 & p & 0 \\ 0 & q & 0 & p \\ 0 & 0 & 0 & 1 \end{array}\right] \end{array}$$

The matrix $\mathbf{Q}$ is given by

$$\mathbf{Q} = \begin{bmatrix} p_{11} & p_{12} \\ p_{21} & p_{22} \end{bmatrix} = \begin{bmatrix} 0 & p \\ q & 0 \end{bmatrix},$$

and

$$\mathbf{R} = \begin{bmatrix} p_{10} & p_{13} \\ p_{20} & p_{23} \end{bmatrix} = \begin{bmatrix} q & 0 \\ 0 & p \end{bmatrix}.$$

Then

$$(\mathbf{I} - \mathbf{Q}) = \begin{bmatrix} 1 & -p \\ -q & 1 \end{bmatrix}.$$

Recall that the inverse of a general 2×2 matrix

$$\mathbf{A} = \begin{bmatrix} a & b \\ c & d \end{bmatrix},$$

is

$$\mathbf{A}^{-1} = \frac{1}{ad - bc} \begin{bmatrix} d & -b \\ -c & a \end{bmatrix}, \qquad ad - bc \neq 0,$$

as can be checked by showing $\mathbf{A}^{-1}\mathbf{A} = \mathbf{A}\mathbf{A}^{-1} = \mathbf{I}$.

Hence the fundamental matrix for this Markov chain is

$$\Phi = \frac{1}{1-pq}\begin{bmatrix} 1 & p \\ q & 1 \end{bmatrix} \tag{8.21}$$

The probabilities of absorption into states 0 and 3 are, by Theorem 8.5,

$$\Pi = \begin{bmatrix} \pi_{10} & \pi_{13} \\ \pi_{20} & \pi_{23} \end{bmatrix} = \Phi R$$

$$= \frac{1}{1-pq}\begin{bmatrix} 1 & p \\ q & 1 \end{bmatrix}\begin{bmatrix} q & 0 \\ 0 & p \end{bmatrix}$$

$$= \frac{1}{1-pq}\begin{bmatrix} q & p^2 \\ q^2 & p \end{bmatrix}$$

In the exercises it is confirmed that the probability of absorption (P_a) at zero for an initial value a, as given by formula (7.17) with $c = 3$, agrees with the values of π_{10} and π_{20}. The row sums of Π are unity, since absorption into one or the other absorbing states is certain. This is also confirmed in Exercise 7.

Example 3

Consider the Markov chain model for the number of A_1 genes but now let the population size be $N = 2$. The state space consists of $0, 1, 2, 3, 4$ and the transition matrix is given by (8.14).

The matrices Q and R are

$$Q = \begin{bmatrix} p_{11} & p_{12} & p_{13} \\ p_{21} & p_{22} & p_{23} \\ p_{31} & p_{32} & p_{33} \end{bmatrix}$$

$$= \tfrac{1}{256}\begin{bmatrix} 108 & 54 & 12 \\ 64 & 96 & 64 \\ 12 & 54 & 108 \end{bmatrix},$$

$$R = \begin{bmatrix} p_{10} & p_{14} \\ p_{20} & p_{24} \\ p_{30} & p_{34} \end{bmatrix}$$

$$= \tfrac{1}{256}\begin{bmatrix} 81 & 1 \\ 16 & 16 \\ 1 & 81 \end{bmatrix}.$$

Thus

$$(I - Q) = \tfrac{1}{256}\begin{bmatrix} 148 & -54 & -12 \\ -64 & 160 & -64 \\ -12 & -54 & 148 \end{bmatrix}.$$

To invert this matrix by hand to find Φ is too messy. However, it will be readily verified (see exercises) that the solutions of the equations

$$(I - Q)\Pi = R,$$

with

$$\Pi = \begin{bmatrix} \pi_{10} & \pi_{14} \\ \pi_{20} & \pi_{24} \\ \pi_{30} & \pi_{34} \end{bmatrix}$$

are

$$\Pi = \tfrac{1}{4} \begin{bmatrix} 3 & 1 \\ 2 & 2 \\ 1 & 3 \end{bmatrix}.$$

In fact the general result with a population of size N is

$$\pi_{k.0} = 1 - \frac{k}{2N}, \qquad \pi_{k.2N} = \frac{k}{2N}, \qquad k = 1, 2, \ldots, 2N - 1, \qquad (8.22)$$

as will also be seen in Exercises 8 and 9.

8.9 THE MEAN TIME TO ABSORPTION

For the Markov chain with transition probabilities given by (8.15) we would like to have as much information as possible concerning the number of time units required to reach an absorbing state from a non-absorbing state. This length of time is of course a random variable which we call the **time to absorption**.

In the population genetics example, the time to absorption of the Markov chain is the time it takes for the heterozygotes to disappear completely from the population. In the random walk with absorbing barriers, the time to absorption is, in the gambling context, the duration of the game or the time required for one player to go broke. In this section we obtain formulas for the mean of the time to absorption.

We define the following two **random variables**.

Definition Let N_{jk} be the number of times the non-absorbing state s_k is occupied until absorption takes place when the Markov chain starts in the non-absorbing state s_j. The collection of N_{jk} forms the $(M - A) \times (M - A)$ matrix N.

Definition Let T_j be the total number of time units until absorption when the Markov chain starts in the non-absorbing state s_j.

The random variable T_j is the time to absorption from state s_j. The collection of T_j, with $j = A + 1, \ldots, M$, forms the $1 \times (M - A)$ row-vector of absorption times for various initial states:

$$\mathbf{T} = [T_{A+1} T_{A+2} \cdots T_M].$$

Since the time to absorption is the total number of times that all the non-absorbing states are occupied, the following relation holds between T_j and the N_{jk}:

$$T_j = \sum_{k=A+1}^{M} N_{jk}. \tag{8.23}$$

The following result gives the expectation of T_j as the sum of the elements in the jth row of the fundamental matrix $\mathbf{\Phi}$.

Theorem 8.6 The mean time to absorption from state s_j is

$$E(T_j) = \sum_{k=A+1}^{M} \phi_{jk} , \qquad j = A+1,\ldots,M, \tag{8.24}$$

where ϕ_{jk} is the (j,k)-element of the fundamental matrix $\mathbf{\Phi}$.
The equations (8.24) may be written in matrix-vector notation as

$$E(\mathbf{T}) = \mathbf{\Phi}\xi$$

where ξ is the $(M-A) \times 1$ column vector

$$\xi = \begin{bmatrix} 1 \\ 1 \\ \cdot \\ \cdot \\ \cdot \\ 1 \end{bmatrix}.$$

Proof The sketch in Fig. 8.3 depicts an initial state s_j and the possible states s_i after the first transition. States $1,\ldots,A$ are absorbing and are lumped together. We will calculate $E(N_{jk})$, there being two separate cases to consider.

Case (i): $k \neq j$.

If the first transition is to an absorbing state, then $N_{jk} = 0$. Hence

$$N_{jk} = 0 \qquad \text{with probability} \sum_{i=1}^{A} p_{ji}.$$

If the first transition is to a non-absorbing state s_i then the total number of times that state s_k is occupied is N_{ik}. Hence

$$N_{jk} = N_{ik} \qquad \text{with probability } p_{ji}, \qquad i = A+1,\ldots,M.$$

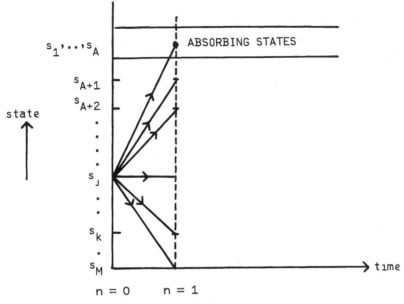

Figure 8.3 Possible transitions from the initial state s_j.

By the law of total probability applied to expectations (see Chapter 1), we must have

$$E(N_{jk}) = \sum_{i=1}^{M} \Pr\{1\text{st transition is from } s_j \text{ to } s_i\}$$

$$\times\, E(N_{jk}|1\text{st transition is from } s_j \text{ to } s_i).$$

The absorbing states contribute zero, so

$$E(N_{jk}) = 0 \times \sum_{i=1}^{A} p_{ji} + \sum_{i=A+1}^{M} p_{ji} E(N_{jk}|1\text{st transition is from } s_j \text{ to } s_i).$$

But we have seen that

$$E(N_{jk}|1\text{st transition is from } s_j \text{ to } s_i) = E(N_{ik}).$$

Hence, for $k \neq j$,

$$E(N_{jk}) = \sum_{i=A+1}^{M} p_{ji} E(N_{ik}). \tag{8.25}$$

Case (ii): $k = j$.

We have $N_{jj} = 1$ if absorption occurs on the first transition, so

$$N_{jj} = 1 \quad \text{with probability } \sum_{i=1}^{A} p_{ji}.$$

If the first transition is to a non-absorbing state s_i, then

$$N_{jj} = 1 + N_{ij} \quad \text{with probability } p_{ji}.$$

Thus,

$$E(N_{jj}) = \sum_{i=1}^{M} \Pr\{\text{1st transition is from } s_j \text{ to } s_i\}$$

$$\times E(N_{jj}|\text{1st transition is from } s_j \text{ to } s_i)$$

$$= 1 \times \sum_{i=1}^{A} p_{ji} + \sum_{i=A+1}^{M} p_{ji}(1 + E(N_{ij}))$$

$$= 1 + \sum_{i=A+1}^{M} p_{ji}E(N_{ij}). \tag{8.26}$$

Introducing the symbol (Kronecker's delta)

$$\delta_{jk} = \begin{cases} 1, & j = k, \\ 0, & j \neq k, \end{cases}$$

equations (8.25) and (8.26) may be summarized as

$$E(N_{jk}) = \delta_{jk} + \sum_{i=A+1}^{M} p_{ji}E(N_{ik}).$$

In matrix form,

$$E(\mathbf{N}) = \mathbf{I} + \mathbf{Q}E(\mathbf{N}),$$

or

$$(\mathbf{I} - \mathbf{Q})E(\mathbf{N}) = \mathbf{I}.$$

Hence

$$E(\mathbf{N}) = \mathbf{\Phi},$$

by definition of $\mathbf{\Phi} = (\mathbf{I} - \mathbf{Q})^{-1}$. Combining this with (8.23),

$$E(T_j) = \sum_{k=A+1}^{M} E(N_{jk}) = \sum_{k=A+1}^{M} \phi_{jk},$$

so the value of $E(T_j)$ is the sum of the elements in the jth row of $\mathbf{\Phi}$ as required.

Example 1

For the Markov chain model exhibiting random drift, when $N = 1$, there is only one non-absorbing state, corresponding to an A_1A_2 individual. The states are s_0, s_1 and s_2. The fundamental matrix is (see (8.20))

$$\mathbf{\Phi} = [2].$$

The expected time to absorption into either state s_0 or s_2 is, by Theorem 8.6, the sum of the elements in the row of the fundamental matrix corresponding to the initial state. Hence the mean time to fixation (A_1A_1 or A_2A_2) is

$$E(T_1) = 2.$$

That is, the mean fixation time is 2 generations.

Example 2

For the random walk with absorbing barriers at 0 and $c = 3$, the fundamental matrix is (see (8.21))

$$\mathbf{\Phi} = \frac{1}{1 - pq} \begin{bmatrix} 1 & p \\ q & 1 \end{bmatrix}.$$

The expected value of T_1, the time to absorption when $X_0 = 1$, is the sum of the elements in the first row of $\mathbf{\Phi}$:

$$E(T_1) = \frac{1 + p}{1 - pq}.$$

Similarly, the expected time to absorption when $X_0 = 2$ is

$$E(T_2) = \frac{1 + q}{1 - pq}.$$

In Exercise 10 it is verified that these results agree with those derived previously (equation (7.23)).

8.10 MUTATION

Genes can be changed by certain stimuli, such as radiation. Sometimes in the 'natural' course of events, chemical accidents may occur which change one allele to another. Such alteration of genetic material is called mutation.

In this section we modify the Markov chain model of Section 8.3 to allow for the possibility of mutation of A_1 alleles to A_2 alleles and vice versa. We will see that this modification drastically alters the properties of the Markov chain.

Suppose that each A_1 allele mutates to A_2 with probability α_1 per generation and that each A_2 allele mutates to A_1 with probability α_2 per generation. By considering the ways in which a gene may be of type A_1 we see that

$$\text{Pr}\{\text{a gene is } A_1 \text{ after mutation}\}$$
$$= \text{Pr}\{\text{gene is } A_1 \text{ before mutation}\}$$
$$\times \text{Pr}\{\text{gene does not mutate from } A_1 \text{ to } A_2\}$$
$$+ \text{Pr}\{\text{gene is } A_2 \text{ before mutation}\}$$
$$\times \text{Pr}\{\text{gene mutates from } A_2 \text{ to } A_1\}.$$

Thus, if there were j genes of type A_1 in the parent population before mutation, the probability p_j of choosing an A_1 gene when we form the next generation is

$$p_j = \frac{j}{2N}(1 - \alpha_1) + \left(1 - \frac{j}{2N}\right)\alpha_2. \tag{8.27}$$

Let X_n be the number of A_1 alleles at generation n. Suppose in fact that there were j genes of type A_1 in generation n. We choose $2N$ genes to form generation $n + 1$ with probability p_j of an A_1 and probability $1 - p_j$ of an A_2 at each trial. Then $\{X_n, n = 0, 1, 2, \ldots\}$ is a temporally homogeneous Markov chain with one-step transition probabilities

$$p_{jk} = \Pr\{X_{n+1} = k \mid X_n = j\}$$
$$= \binom{2N}{k}(p_j)^k(1 - p_j)^{2N-k}; \qquad j, k = 0, 1, 2, \ldots, 2N, \tag{8.28}$$

with p_j given by (8.27).

Example

Let $N = 1$, $\alpha_1 = \alpha_2 = \frac{1}{4}$. Then, substituting in (8.27),

$$p_j = \frac{j}{4} + \frac{1}{4}.$$

This gives $p_0 = \frac{1}{4}, p_1 = \frac{1}{2}, p_2 = \frac{3}{4}$. The elements of $\mathbf{P}$ are, from (8.28), for $k = 0, 1, 2$,

$$p_{0k} = \binom{2}{k}(\tfrac{1}{4})^k(\tfrac{3}{4})^{2-k}$$

$$p_{1k} = \binom{2}{k}(\tfrac{1}{2})^2$$

$$p_{2k} = \binom{2}{k}(\tfrac{3}{4})^k(\tfrac{1}{4})^{2-k}.$$

Evaluating these we obtain

$$\mathbf{P} = \tfrac{1}{16}\begin{bmatrix} 9 & 6 & 1 \\ 4 & 8 & 4 \\ 1 & 6 & 9 \end{bmatrix}.$$

There are no absorbing states, as there are no ones on the principal diagonal. All the elements of $\mathbf{P}$ are non-zero, so at any time step a transition is possible from any state to any other. We will see that in contrast to the case where there is no mutation, an equilibrium probability distribution is eventually achieved.

8.11 STATIONARY DISTRIBUTIONS

Let **P** be the transition matrix of a temporally homogeneous Markov chain $\{X_n, n = 0, 1, 2, \ldots\}$. Suppose there exists a **probability vector** $\hat{p}$ (i.e., a row vector with non-negative components whose sum is unity) satisfying

$$\hat{p}P = \hat{p}. \tag{8.29}$$

Now let the probability distribution of X_n be given by $p(n)$ and suppose the process starts with

$$p(0) = \hat{p}.$$

Then we must have

$$p(1) = p(0)P = \hat{p}P = \hat{p}$$
$$p(2) = p(1)P = \hat{p}P = \hat{p},$$

and it can be seen that

$$p(n) = \hat{p},$$

for all $n = 0, 1, 2, \ldots$

We find that if the initial probability distribution is given by $\hat{p}$, the probability distribution of the process at each time step is the same. We call such a probability distribution **stationary** or **time-invariant**. The random variables $X_0, X_1, X_2, \ldots$ are thus identically distributed if X_0 has the distribution $\hat{p}$.

Definition Let **P** be the transition matrix of a temporally homogeneous Markov chain. If there exists a probability vector $\hat{p}$ such that $\hat{p} = \hat{p}P$, then $\hat{p}$ is called a stationary distribution for the Markov chain.

Note on terminology

A vector x is said to be a (left) **eigenvector** of the matrix A if xA is a scalar (real or complex) multiple of x. That is,

$$xA = \lambda x,$$

where λ is a scalar called the corresponding **eigenvalue**. According to (8.29) $\hat{p}P = 1\hat{p}$. Hence a stationary distribution is an eigenvector of **P** with eigenvalue 1. Any non-zero multiple of an eigenvector is also an eigenvector with the same eigenvalue. To fix $\hat{p}$ we insist that its components sum to unity.

Example

Let the transition matrix of a temporally homogeneous Markov chain be

$$P = \begin{bmatrix} 0.4 & 0.6 \\ 0.2 & 0.8 \end{bmatrix}. \tag{8.30}$$

An eigenvector $\mathbf{x} = [x_1 x_2]$ of $\mathbf{P}$ with eigenvalue 1 must satisfy

$$[x_1 \quad x_2]\begin{bmatrix} 0.4 & 0.6 \\ 0.2 & 0.8 \end{bmatrix} = [x_1 \quad x_2].$$

Thus

$$0.4x_1 + 0.2x_2 = x_1$$
$$0.6x_1 + 0.8x_2 = x_2.$$

From the first (or second) of these equations we find

$$x_2 = 3x_1.$$

Hence any multiple of the row vector

$$\mathbf{x} = [1 \quad 3]$$

is an eigenvector with eigenvalue 1. To obtain a probability vector we must divide by the sum of the components. Thus a stationary probability vector for this Markov chain is

$$\hat{\mathbf{p}} = [0.25 \quad 0.75] \doteq [\hat{p}_1 \quad \hat{p}_2].$$

8.12 APPROACH TO A STATIONARY DISTRIBUTION AS $n \to \infty$

Consider again the Markov chain with $\mathbf{P}$ given by (8.30). Computing successive powers of $\mathbf{P}$ we find

$$\mathbf{P}^2 = \begin{bmatrix} 0.28 & 0.72 \\ 0.24 & 0.76 \end{bmatrix}$$

$$\mathbf{P}^3 = \begin{bmatrix} 0.256 & 0.744 \\ 0.248 & 0.752 \end{bmatrix}$$

$$\mathbf{P}^4 = \begin{bmatrix} 0.2512 & 0.7488 \\ 0.2496 & 0.7504 \end{bmatrix}$$

It would seem, and we will see that it is true, that as n increases $\mathbf{P}^n$ is approaching, element by element, the matrix

$$\hat{\mathbf{P}} = \begin{bmatrix} 0.25 & 0.75 \\ 0.25 & 0.75 \end{bmatrix}.$$

That is,

$$\lim_{n \to \infty} \mathbf{P}^n = \hat{\mathbf{P}}. \tag{8.31}$$

Note that each row of $\hat{\mathbf{P}}$ is the same as the stationary probability vector $\hat{\mathbf{p}}$, so

$$\hat{\mathbf{P}} = \begin{bmatrix} \hat{p}_1 & \hat{p}_2 \\ \hat{p}_1 & \hat{p}_2 \end{bmatrix}.$$

In terms of matrix elements,

$$\lim_{n \to \infty} p_{jk}^{(n)} = \hat{p}_k,$$

regardless of the value of j.

Let us see what happens, if (8.31) is true, to the probability distribution of X_n as $n \to \infty$, for an arbitrary initial distribution

$$\mathbf{p}(0) = [p_1(0) \quad p_2(0)].$$

Since

$$\mathbf{p}(n) = \mathbf{p}(0)\mathbf{P}^n,$$

we have

$$\lim_{n \to \infty} \mathbf{p}(n) = \mathbf{p}(0)\hat{\mathbf{P}}$$

$$= [p_1(0) \quad p_2(0)] \begin{bmatrix} \hat{p}_1 & \hat{p}_2 \\ \hat{p}_1 & \hat{p}_2 \end{bmatrix}$$

$$= [\hat{p}_1(p_1(0) + p_2(0)) \quad \hat{p}_2(p_1(0) + p_2(0))]$$

$$= [\hat{p}_1 \quad \hat{p}_2],$$

since the components of $\mathbf{p}(0)$ must add to one. Thus

$$\mathbf{p}(n) \xrightarrow[n \to \infty]{} \hat{\mathbf{p}}$$

for an arbitrary initial probability distribution. Under these circumstances we say that the distribution of X_n approaches a **steady-state distribution** which coincides with the stationary probability vector $\hat{\mathbf{p}}$.

The question arises as to what conditions guarantee the approach to a steady-state distribution. Before stating the main result we make the following definition.

Definition A Markov chain is regular if there is a finite positive integer m such that after m time-steps, every state has a non-zero chance of being occupied, no matter what the initial state.

Notation

If every element a_{jk} of a matrix $\mathbf{A}$ satisfies the inequality

$$a_{jk} > 0$$

then we write

$$\mathbf{A} > 0.$$

Thus, for a regular Markov chain with transition matrix $\mathbf{P}$, there exists an $m > 0$ such that

$$\mathbf{P}^m > 0.$$

In Exercise 16 it is proved that

$$P^m > 0 \Rightarrow P^{m+k} > 0, \qquad k = 1, 2, \ldots$$

For regular Markov chains we have the following result concerning steady-state distributions.

Theorem 8.7 Let $X = \{X_0, X_1, \ldots\}$ be a regular temporally homogeneous Markov chain with a finite number M of states and transition matrix P. Then,

(i) Regardless of the value of $j = 1, 2, \ldots, M$,

$$\lim_{n \to \infty} p_{jk}^{(n)} = \hat{p}_k, \qquad k = 1, 2, \ldots, M.$$

or equivalently,

(ii) $$\lim_{n \to \infty} P^n = \hat{P},$$

where $\hat{P}$ is a matrix whose rows are identical and equal to the probability vector

$$\hat{p} = [\hat{p}_1 \hat{p}_2 \cdots \hat{p}_M].$$

(iii) No matter what the probability distribution p(0) of X_0, the probability distribution of X_n approaches $\hat{p}$ as $n \to \infty$:

$$p(n) \xrightarrow[n \to \infty]{} \hat{p}$$

(iv) $\hat{p}$ is the unique solution of

$$\hat{p}P = \hat{p}$$

satisfying $\hat{p} > 0$ and $\sum_k \hat{p}_k = 1$.

For a proof see Kemeny and Snell (1960) or Feller (1968). Note that in the terminology of Feller a regular Markov chain is irreducible, aperiodic and has only ergodic states. The terminology for Markov chains is confusing as different authors use the same word with different meanings as well as several different words for the same thing. It seemed best to avoid these altogether in an introductory treatment. A matrix A satisfying $A^m > 0$ for some positive integer m is called **primitive**. The theory of such matrices is well developed, including the useful Perron-Frobenius theorems. See, for example, Seneta (1983).

Example 1

For the population genetics example of the previous section

$$P = \tfrac{1}{16} \begin{bmatrix} 9 & 6 & 1 \\ 4 & 8 & 4 \\ 1 & 6 & 9 \end{bmatrix}.$$

Since $\mathbf{P} > 0$ we see from Theorem 8.7 that a steady-state probability distribution will be approached as $n \to \infty$. To obtain the stationary distribution we
must find a left eigenvector of $\mathbf{P}$ with eigenvalue 1 whose components add to
unity.

Any eigenvector $\mathbf{x} = [x_1 \quad x_2 \quad x_3]$ with eigenvalue 1 must satisfy $\mathbf{xP} = \mathbf{x}$.
Hence

$$[x_1 \quad x_2 \quad x_3] \begin{bmatrix} 9 & 6 & 1 \\ 4 & 8 & 4 \\ 1 & 6 & 9 \end{bmatrix} = 16[x_1 \quad x_2 \quad x_3].$$

This yields three equations, of which only two are needed. The first two
equations are

$$9x_1 + 4x_2 + x_3 = 16x_1$$
$$6x_1 + 8x_2 + 6x_3 = 16x_2,$$

or

$$7x_1 - 4x_2 - x_3 = 0$$
$$-6x_1 + 8x_2 - 6x_3 = 0.$$

Since one of the components of $\mathbf{x}$ is arbitrary we may set $x_3 = 1$ and solve

$$7x_1 - 4x_2 = 1$$
$$-6x_1 + 8x_2 = 6.$$

This yields $x_1 = 1, x_2 = 3/2$ and $x_3 = 1$ so any left eigenvector of $\mathbf{P}$ with
eigenvalue 1 is a non-zero multiple of

$$\mathbf{x} = [1 \quad 3/2 \quad 1].$$

The sum of the components of x is 7/2 so dividing x by 7/2 we obtain the
required stationary probability vector

$$\hat{\mathbf{p}} = [2/7 \quad 3/7 \quad 2/7].$$

For any initial probability vector $\mathbf{p}(0)$, the probability distribution of X_n
approaches $\hat{\mathbf{p}}$. In particular, even if the population starts with say, all $A_1 A_1$, so

$$\mathbf{p}(0) = [0 \quad 0 \quad 1],$$

there is probability 3/7 that the population will eventually be heterozygous.
Compare this behaviour with random drift.

Example 2

This example, from Ash (1970), shows that a stationary distribution may exist
but this does not imply that a steady-state is approached as $n \to \infty$.

Consider a Markov chain with two states and

$$\mathbf{P} = \begin{bmatrix} 0 & 1 \\ 1 & 0 \end{bmatrix}$$

so that transitions are only possible from one state to the other. Solving

$$\mathbf{xP} = \mathbf{x}$$

or,

$$[x_1 \quad x_2]\begin{bmatrix} 0 & 1 \\ 1 & 0 \end{bmatrix} = [x_1 \quad x_2]$$

gives $x_2 = x_1$. Hence

$$\hat{\mathbf{p}} = [\tfrac{1}{2} \quad \tfrac{1}{2}],$$

is a stationary probability vector. However, as $n \to \infty$, $\mathbf{P}^n$ does not approach a constant matrix because

$$\mathbf{P}^n = \begin{cases} \begin{bmatrix} 0 & 1 \\ 1 & 0 \end{bmatrix}, & n = 1, 3, 5, \ldots \\ \begin{bmatrix} 1 & 0 \\ 0 & 1 \end{bmatrix}, & n = 2, 4, 6, \ldots \end{cases}$$

The conditions of Theorem 8.7 are violated, this not being a regular Markov chain. It is seen that state 2 can only be entered from state 1 and vice versa on time steps $1, 3, 5, \ldots$. Such a Markov chain is called **periodic** or **cyclic** with period 2. For a discussion of such Markov chains see Feller (1968).

REFERENCES

Ash, R.B. (1970). *Basic Probability Theory*. Wiley, New York.
Bailey, N.T.J. (1964). *The Elements of Stochastic Processes*. Wiley, New York.
Crow, J.F. and Kimura, M. (1970). *An Introduction to Population Genetics Theory*. Harper and Row, New York.
Ewens, W.J. (1979). *Mathematical Population Genetics*. Springer-Verlag, New York.
Feller, W. (1968). *An Introduction to Probability Theory and its Applications*. Wiley, New York.
Isaacson, D.L. and Madsen, R.W. (1976). *Markov Chains Theory and Applications*. Wiley, New York.
Kemeny, J.G. and Snell, J.L. (1960). *Finite Markov Chains*. Van Nostrand, Princeton, N.J.
Seneta, E. (1983). *Nonnegative Matrices and Markov Chains*, Springer-Verlag, New York.
Strickberger, M.W. (1968). *Genetics*. Macmillan, New York.
Watson, J.D. (1970). *Molecular Biology of the Gene*, Benjamin, New York.

Additional works including advanced and specialized treatments not referenced in the text

Bartholomew, D.J. (1967). *Stochastic Models for Social Processes*. Wiley, London.
Chung, K.L. (1967). *Markov chains with Stationary Transition Probabilities*. Springer-Verlag, New York.

Cox, D.R. and Miller, H.D. (1965). *The Theory of Stochastic Processes*. Wiley, New York.

Karlin, S. and Taylor, H.M. (1975). *A First Course in Stochastic Processes*. Academic Press, New York.

Kemeny, J.G., Snell, J.L. and Knaff, A.W. (1966). *Denumerable Markov Chains*. Van Nostrand, Princeton, N.J.

Parzen, E. (1962). *Stochastic Processes*. Holden-Day, San Francisco.

EXERCISES

1. A gene is present in human populations which has two alleles A_1 and A_2. If a group initially has 40 $A_1 A_1$, 30 $A_1 A_2$ or $A_2 A_1$ and 30 $A_2 A_2$ individuals, what will the equilibrium (HW) genotype frequencies be?

2. Complete the proof of Theorem 8.2; that

$$\Pr(X_0 = s_{j_0}, X_1 = s_{j_1}, \ldots, X_n = s_{j_n}) = P_{j_0}(0) p_{j_0 j_1}(1) \cdots p_{i_{n-1} j_n}(n)$$

   for $n \geqslant 1$. (*Hint*: Use mathematical induction.)

3. Establish the **Chapman–Kolmogorov equations**

$$p_{jk}^{(m+n)} = \sum_{i=1}^{M} p_{ji}^{(m)} p_{ik}^{(n)}.$$

   (*Hint*: Use matrix multiplication.)

4. Show that the matrix with elements given by (8.13) is stochastic.

5. Any stochastic matrix defines a temporally homogeneous Markov chain. Which of the following matrices are stochastic?

   (a) $\begin{bmatrix} 1/4 & 3/4 \\ 1 & 0 \end{bmatrix}$ (b) $\begin{bmatrix} 1/2 & 1/2 & 0 \\ 1/2 & 1/4 & 0 \\ 1 & 0 & 0 \end{bmatrix}$

   (c) $\begin{bmatrix} 0 & 1 & 0 & 0 \\ 1/3 & 1/3 & 0 & 1/3 \\ 1/2 & 1/4 & 1/8 & 1/8 \\ 1/2 & 0 & 1/4 & 1/4 \end{bmatrix}$

6. For the Markov chain for random mating with no mutation, the transition matrix when $N = 1$ is

$$P = \begin{bmatrix} 1 & 0 & 0 \\ 1/4 & 1/2 & 1/4 \\ 0 & 0 & 1 \end{bmatrix}$$

   If X_0 has the distribution $\mathbf{p}(0) = [0 \quad \tfrac{1}{2} \quad \tfrac{1}{2}]$, find the probability distributions of X_1, X_2, X_3 and X_4. Plot these distributions and observe the phenomenon of random drift.

7. The matrix Π of absorption probabilities for the simple random walk with absorbing barriers at 0 and 3 was found to be

$$\Pi = \frac{1}{1-pq}\begin{bmatrix} q & p^2 \\ q^2 & p \end{bmatrix}.$$

Verify that
(a) the row sums of Π are unity
(b) the probabilities of absorption at 0 agree with those given by (7.17).

8. For the genetic Markov chain (Section 8.3) with a population of N diploid individuals, find the matrices $\mathbf{Q}$ and $\mathbf{R}$. Verify that the matrix Π of absorption probabilities

$$\Pi = \begin{bmatrix} 1 - 1/2N & 1/2N \\ 1 - 2/2N & 2/2N \\ \vdots & \vdots \\ 1 - k/2N & k/2N \\ \vdots & \vdots \\ 1/2N & 1 - 1/2N \end{bmatrix}$$

satisfies

$$(\mathbf{I} - \mathbf{Q})\Pi = \mathbf{R}.$$

9. Prove that the Markov chain $\{X_n\}$ for random genetic drift defined in Section 8.3 is a martingale. (cf. Exercise 14 of Chapter 7.) Use the optional stopping theorem to deduce immediately that the probabilities of fixation are given by (8.22).

10. For the simple random walk with absorbing barriers at 0 and 3, verify that the formulas

$$E(T_1) = \frac{1+p}{1-pq}, \quad E(T_2) = \frac{1+q}{1-pq},$$

for the expected times to absorption from $X_0 = 1$, $X_0 = 2$, respectively, agree with those given by (7.23).

11. The following problem is based upon one in Kemeny and Snell (1960). In each year of a three-year degree course, a university student has probability p of not returning the following year, probability q of having to repeat the year and probability r of passing ($p + q + r = 1$). The states are: dropped out (s_1), graduated (s_2), is a third-year student (s_3), is a second-year student (s_4), and is a first-year student (s_5). Find the transition matrix $\mathbf{P}$ and the matrices $\mathbf{Q}$ and $\mathbf{R}$. (Note that this is a random walk with absorbing barriers.)

12. For the Markov chain of Exercise 11, solve the equations $(\mathbf{I} - \mathbf{Q})\mathbf{\Phi} = \mathbf{I}$ to obtain the fundamental matrix $\mathbf{\Phi} = (\mathbf{I} - \mathbf{Q})^{-1}$.

13. For the Markov chain of Exercise 11, find a student's chances of graduating if he is in years 1, 2 and 3.

14. For the Markov chain of Exercise 11, find the average number of years a first-year, second-year and third-year student will remain in university.

15. The following example is based on an application discussed in Isaacson and Madsen (1976). Farms are divided into four categories: very small (s_1), very large (s_2), large (s_3) and small (s_4). Farms grow or shrink as land is bought or sold. It is assumed that once a farm is very small or very large it stays as such. Small and large farms increase in size each year into the next category with probability $\frac{1}{2}$, remain the same size with probability $\frac{1}{4}$ and decrease in size to the category below with probability $\frac{1}{4}$. Find the transition matrix and the expected time for a small farm to become either very small or very large.

16. Prove that $\mathbf{P}^m > 0 \Rightarrow \mathbf{P}^{m+k} > 0$, $k = 1, 2, \ldots$

17. The following learning model, due to Bush and Mosteller, is discussed in Bailey (1964). In a learning experiment let s_1 be a correct response and s_2 an incorrect response. The response at any trial depends only on the result of the previous trial and the transition matrix is

$$\mathbf{P} = \begin{bmatrix} 1-p & p \\ q & 1-q \end{bmatrix}, \qquad 0 < p, q < 1.$$

Let X_n be the response on trial n, $n = 0, 1, 2, \ldots$

(a) Find the stationary probability vector $\hat{\mathbf{p}}$.

(b) Will the probability distribution of X_n approach $\hat{\mathbf{p}}$ as $n \to \infty$?

(c) Find the matrix $\hat{\mathbf{P}}$.

(d) Prove, using induction or otherwise, that

$$\mathbf{P}^n = \frac{1}{p+q} \begin{bmatrix} q & p \\ q & p \end{bmatrix} + \frac{(1-p-q)^n}{p+q} \begin{bmatrix} p & -p \\ -q & q \end{bmatrix}.$$

Hence verify your result (c).

(e) If the initial response is correct so $\mathbf{p}(0) = \begin{bmatrix} 1 & 0 \end{bmatrix}$, what is the probability of a correct response on trial n?

18. For a simple random walk assume there are reflecting barriers at 0 and 3. That is, when the particle gets to 0 or 3 it goes on the next step to states 1 or 2 (respectively) with probability one. Thus the transition matrix is

$$\mathbf{P} = \begin{bmatrix} 0 & 1 & 0 & 0 \\ q & 0 & p & 0 \\ 0 & q & 0 & p \\ 0 & 0 & 1 & 0 \end{bmatrix}$$

    (a) Without doing any calculations, is this a regular Markov chain? Why?

    (b) If the answer to (a) is yes, compute the equilibrium probability distribution $\hat{\mathbf{p}}$.

    (c) If $X_0 = 3$, what is the eventual probability that the position of the particle is 3?

19. In the Markov chain model of random mating with mutation in a population of size N, find $\mathbf{P}$ if $\alpha_1 = \alpha_2 = \alpha \neq 0$. Given an arbitrary initial probability distribution $\mathbf{p}(0)$, find $\mathbf{p}(1)$ and deduce that the stationary distribution is attained in one generation.

20. What will happen in the Markov chain model of random mating with mutation if $\alpha_1 \neq 0$ but $\alpha_2 = 0$?

9

Population growth I: birth and death processes

9.1 INTRODUCTION

It is clearly desirable that governments and some businesses be able to predict future human population numbers. Not only are the total numbers of male and female individuals of interest but also the numbers in certain categories such as age groups. The subject which deals with population numbers and movements is called **demography**.

Some of the data of concern to human demographers is obtained from our filling out census forms. The type of data is exemplified by that in Tables 9.1 and 9.2. In Table 9.1 is given the total population of Australia at various times since 1881 and Table 9.2 contains some data on births and deaths and their rates. Notice the drastic fall in the birth rate in the last few decades compared with an almost steady death rate.

Table 9.1

Time	Population of Australia (thousands)*
3 April 1881	2 250.2
5 April 1891	3 177.8
31 March 1901	3 773.8
3 April 1911	4 455.0
4 April 1921	5 435.7
30 June 1933	6 629.8
30 June 1947	7 579.4
30 June 1954	8 986.5
30 June 1961	10 548.3
30 June 1966	11 599.5
30 June 1971	12 937.2
30 June 1979	14 417.2
30 June 1981	14 923.3

*Obtained from Cameron (1982) and Australian Bureau of Statistics.

Table 9.2

| Time period | Annual (average) number | | Annual rates per thousand | |
	Births	Deaths	Births	Deaths
1956–60	222459	86488	22.59	8.78
1961–65	232952	95465	21.34	8.75
1966–70	240325	107263	19.95	8.90
1971–75	253438	111216	18.99	8.32
1976	227810	112662	16.37	8.10
1977	226291	108790	16.08	7.73
1978	224181	108425	15.73	7.61

Deterministic model

An accurate mathematical model for the growth of a population would be a very useful thing to have. There is a substantial literature on various models, as exemplified by the books of Bartlett (1960), Keyfitz (1968), Pollard (1973) and Ludwig (1978). A first division of such models is into **deterministic** versus **stochastic** ones. In the former category there are no chance effects.

Let $N(t)$ be the population size at time $t \geqslant 0$ and assume that the initial population size $N(0) > 0$ is given. The number of individuals at t is a non-negative integer, but it is convenient to assume that $N(t)$ is a differentiable function of time. A simple differential equation for N is obtained by letting there be $b\Delta t$ births and $d\Delta t$ deaths per individual in $(t, t + \Delta t]$ where $b, d \geqslant 0$ are the **per capita birth and death rates**. Then

$$N(t + \Delta t) = N(t) + N(t)b\Delta t - N(t)d\Delta t$$

and it follows, upon rearranging and taking the limit $\Delta t \to 0$, that

$$\frac{dN}{dt} = (b - d)N, \qquad t > 0.$$

This differential equation is called the **Malthusian growth law** (after Thomas Malthus, whose essay on population appeared in 1798). Its solution is

$$\boxed{N(t) = N(0)e^{(b-d)t}},$$

and once we specify $N(0)$ the population size $N(t)$ is determined for all $t > 0$. Three qualitatively different behaviours are possible, depending on the relative magnitudes of b and d. These are illustrated in Fig. 9.1.

It is seen that

$$\lim_{t \to \infty} N(t) = \begin{cases} 0, & b < d \text{ (exponential decay)} \\ N(0), & b = d \text{ (constant population)} \\ \infty, & b > d \text{ (exponential growth)}. \end{cases}$$

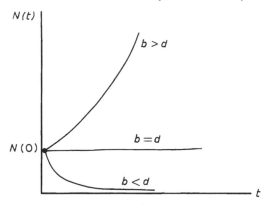

Figure 9.1 Population size as a function of time for various relative magnitudes of the birth and death rates in the Malthusian growth model.

Experience tells us that the times of occurrence of births and deaths are not predictable in real populations. Hence the population sizes $N = \{N(t), t \geq 0\}$ constitute a random process. As observed above, the number of individuals is a non-negative integer so N is a random process with a **discrete state space** $\{0, 1, 2, \ldots\}$ and a continuous parameter set $\{t \geq 0\}$. The models we will consider do not contain very many elements of reality. They are nevertheless interesting because they may be solved exactly and provide a starting point for more complicated models.

9.2 SIMPLE POISSON PROCESSES

The random processes discussed in this section are of fundamental importance even though, as will be seen, their use as population growth models is very limited. The following process has a different parameterization from that defined in Chapter 3.

Definition A collection of random variables $N = \{N(t), t \geq 0\}$ is called a simple Poisson process with intensity λ if the following hold:

  (i) **$N(0) = 0$.**
 (ii) **For any collection of times $0 \leq t_0 < t_1 < \ldots < t_{n-1} < t_n < \infty$ the random variables $N(t_k) - N(t_{k-1})$, $k = 1, 2, \ldots, n$, are mutually independent.**
(iii) **For any pair of times $0 \leq t_1 < t_2$, the random variable $N(t_2) - N(t_1)$ is Poisson distributed with parameter $\lambda(t_2 - t_1)$.**

Condition (i) is a convenient initialization. The quantity $\Delta N_k = N(t_k) - N(t_{k-1})$ is the change or increment in the process in the time interval

$(t_{k-1}, t_k]$. Condition (ii) says that the increments in disjoint (non-overlapping) time intervals are independent. Condition (iii) gives the probability law of the increments. Then

$$\Pr\{N(t_2) - N(t_1) = k\} = \frac{(\lambda(t_2 - t_1))^k e^{-\lambda(t_2 - t_1)}}{k!}, \qquad k = 0, 1, 2, \ldots$$

and in particular, choosing $t_1 = 0$ and $t_2 = t > 0$ we have, since $N(0) = 0$,

$$\boxed{\Pr\{N(t) = k\} = \frac{(\lambda t)^k e^{-\lambda t}}{k!}}, \qquad k = 0, 1, 2, \ldots \qquad (9.1)$$

The mean and variance of $N(t)$ are therefore

$$E(N(t)) = \mathrm{Var}(N(t)) = \lambda t.$$

Sample paths

What does a typical realization of a simple Poisson process look like? To get some insight, let Δt be a very small time increment and consider the increment

$$\Delta N(t) = N(t + \Delta t) - N(t).$$

The probability law of this increment is

$$\Pr\{\Delta N(t) = k\} = \frac{(\lambda \Delta t)^k e^{-\lambda \Delta t}}{k!}$$

$$= \begin{cases} 1 - \lambda \Delta t + o(\Delta t), & k = 0 \\ \lambda \Delta t + o(\Delta t), & k = 1 \\ o(\Delta t), & k \geqslant 2. \end{cases} \qquad (9.2)$$

We see therefore that when Δt is very small, $N(t + \Delta t)$ is most likely to be the same as $N(t)$, with probability $\lambda \Delta t$ that it is one larger, and there is a negligible chance that it differs by more than one from $N(t)$. We may conclude that sample paths are right-continuous step functions with discontinuities of magnitude unity - see Fig. 9.2. Note that equation (9.2) can be used as a definition and the probability law of the increments derived from it - see Exercise 2.

The simple Poisson process is called **simple** because all of its jumps have the same magnitude - unity in the above case. This contrasts with **compound Poisson processes** in which jumps may be any of several magnitudes (see for example Parzen, 1962).

To relate the Poisson process to a growth model we imagine that each time a new individual is born, $N(t)$ jumps up by unity. Thus $N(t)$ records the number of births in $(0, t]$, i.e. up to and including time t and $\{N(t), t \geqslant 0\}$ may be regarded as a birth process. If in Fig. 9.2 we place a cross on the t-axis at each

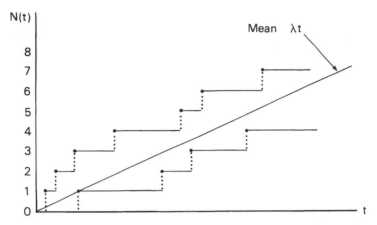

Figure 9.2 Two representative sample paths for a simple Poisson process. The **mean value function** λt is also indicated.

jump of $N(t)$ we obtain a collection of points. Thus there is a close relationship between the simple Poisson processes of this section and the Poisson point processes of Section 3.4. This is elaborated on in the exercises.

Poisson processes, though clearly limited as population growth models, find many other applications. Some were mentioned in Chapter 3. Furthermore they may result when several sparse point processes, not themselves Poisson, are pooled together. This makes them useful in diverse areas (Tuckwell, 1981).

9.3 MARKOV CHAINS IN CONTINUOUS TIME

In Chapter 8 we encountered Markov random processes which take on discrete values and have a discrete parameter set. Markov processes of that kind are called Markov chains. However, Markov processes in **continuous time** and with discrete state space are also called Markov chains. Usually we mention that they have continuous parameter set when we refer to them in order to distinguish them from the former kind of process. If $X = \{X(t), t \geq 0\}$ is such a process, the Markov property may be written, with $t_0 < t_1 < t_2 \cdots < t_{n-1} < t_n$,

$$\Pr\{X(t_n) = s_n | X(t_{n-1}) = s_{n-1}, \; X(t_{n-2}) = s_{n-2}, \ldots, X(t_1) = s_1, X(t_0) = s_0\}$$
$$= \Pr\{X(t_n) = s_n | X(t_{n-1}) = s_{n-1}\},$$

where $\{s_0, s_1, \ldots, s_n\}$ is any set in the state space of the process. The quantities

$$p(s_n, t_n | s_{n-1}, t_{n-1}) = \Pr\{X(t_n) = s_n | X(t_{n-1}) = s_{n-1}\}, 0 \leq t_{n-1} < t_n < \infty, \quad (9.3)$$

are called the **transition probabilities** of the continuous time Markov chain. If the transition probabilities depend only on time **differences** they are said to be **stationary** and the corresponding process is called **temporally homogeneous**.

Simple Poisson processes are examples of continuous time Markov chains with stationary transition probabilities which may be written

$$p(k, t \,|\, j, s) = \Pr\{N(t) = k \,|\, N(s) = j\}$$
$$= \frac{(\lambda(t - s))^{k - j} e^{-\lambda(t - s)}}{(k - j)!}, \quad k - j = 0, 1, 2, \dots$$

where $0 \leqslant s < t$ and $j \leqslant k$ are two non-negative integers. Further examples are discussed in the following sections.

9.4 THE YULE PROCESS

Among the deficiencies of the Poisson process as a model for the growth of populations is that no matter how large the existing population, the chance of a birth in any time interval is always the same. The **simple birth process** we are about to describe was proposed by Yule (1924) as a model for the appearance of new species. Its applicability as a population growth model is limited except perhaps in a few cases such as algae undergoing relatively unchecked reproduction in large lakes (Pielou, 1969). The Yule process is sometimes called the **Yule–Furry Process** due to a related application in physics by Furry (1937). Again we denote by $N(t)$ the number of individuals in existence at time t.

Assumptions on births to individuals

We begin with the following assumptions concerning the births which occur to the individual members of the population.

(i) Births occur to any individual independently of those to any other individual.

(ii) In any small time interval of length Δt, the probability of an offspring to any individual is $\lambda \Delta t + o(\Delta t)$, the probability of no offspring is $1 - \lambda \Delta t + o(\Delta t)$ and the probability of more than one offspring is $o(\Delta t)$.

Notice that there are no deaths – individuals persist indefinitely and are forever capable of producing offspring.

Population birth probabilities

Suppose that there are known to be n individuals at time t so that $N(t) = n$. Under the above assumptions (i) and (ii), the number of births in the whole population in $(t, t + \Delta t]$ is a binomial random variable with parameters n and $\lambda \Delta t$. We can drop reference to $o(\Delta t)$ as such terms eventually make no contribution. Then

$$\Pr\{k \text{ births in } (t, t + \Delta t] \mid N(t) = n\}$$

$$= \binom{n}{k}(\lambda\Delta t)^k(1 - \lambda\Delta t)^{n-k}, \qquad k = 0, 1, \ldots, n.$$

If $k = 0$ we have

$$\Pr\{0 \text{ births in } (t, t + \Delta t] \mid N(t) = n\} = (1 - \lambda\Delta t)^n$$
$$= 1 - \lambda n\Delta t + o(\Delta t), \qquad (9.4)$$

whereas for $k = 1$,

$$\Pr\{1 \text{ birth in } (t, t + \Delta t] \mid N(t) = n\} = \lambda n\Delta t(1 - \lambda\Delta t)^{n-1}$$
$$= \lambda n\Delta t + o(\Delta t). \qquad (9.5)$$

Also, for $k \geqslant 2$ we find

$$\Pr\{k \text{ births in } (t, t + \Delta t] \mid N(t) = n\} = o(\Delta t). \qquad (9.6)$$

Differential-difference equations satisfied by the transition probabilities

We consider a continuous time Markov chain $N = \{N(t), t \geqslant 0\}$ with initial value

$$N(0) = n_0 > 0,$$

subject to the evolutionary laws (9.4)–(9.6). Define the transition probabilities

$$p_n(t) = \Pr\{N(t) = n \mid N(0) = n_0\}, \qquad n_0 > 0, \ t > 0,$$

and note that these are stationary. Our aim is first to find equations governing the evolution in time of p_n and then to solve them.

To obtain a differential equation for p_n we seek a relation between $p_n(t)$ and $p_n(t + \Delta t)$. If $N(t + \Delta t) = n > n_0$ then, ignoring the possibility of more than one birth, we must have

$$N(t) = \begin{cases} n \text{ and no births in } (t, t + \Delta t], \\ n - 1 \text{ and one birth in } (t, t + \Delta t]. \end{cases}$$

Dropping reference to the initial state, the law of total probability gives

$$\Pr\{N(t + \Delta t) = n\} = \Pr\{N(t + \Delta t) = n \mid N(t) = n\}$$
$$\times \Pr\{N(t) = n\} + \Pr\{N(t + \Delta t) = n \mid N(t) = n - 1\}$$
$$\times \Pr\{N(t) = n - 1\}. \qquad (9.7)$$

In symbols this is written, using (9.4) and (9.5),

$$p_n(t + \Delta t) = (1 - \lambda n\Delta t)p_n(t) + \lambda(n - 1)\Delta t p_{n-1}(t) + o(\Delta t), \quad n > n_0.$$

This rearranges to

$$\frac{p_n(t + \Delta t) - p_n(t)}{\Delta t} = \lambda[(n - 1)p_{n-1}(t) - np_n(t)] + \frac{o(\Delta t)}{\Delta t}.$$

Taking the limit $\Delta t \to 0$ we obtain the required equation

$$\boxed{\frac{dp_n}{dt} = \lambda[(n - 1)p_{n-1} - np_n]} , \qquad n = n_0 + 1, n_0 + 2, \ldots \qquad (9.8)$$

since $o(\Delta t)/\Delta t \to 0$ as $\Delta t \to 0$ by definition. When $n = n_0$ there is no possibility that the population was $n_0 - 1$ so (9.7) becomes

$$\Pr\{N(t + \Delta t) = n_0\} = \Pr\{N(t + \Delta t) = n_0 | N(t) = n_0\} \Pr\{N(t) = n_0\}$$

which leads to

$$\boxed{\frac{dp_{n_0}}{dt} = -\lambda n_0 p_{n_0}} . \qquad (9.9)$$

Initial conditions

Equations (9.8) and (9.9) are first-order differential equations in time. To solve them the values of $p_n(0)$ are needed and since an initial population of n_0 individuals was assumed, we have

$$p_n(0) = \begin{cases} 1, & n = n_0, \\ 0, & n > n_0. \end{cases}$$

Solutions of the differential-difference equations

The solution of (9.9) with initial value unity is

$$p_{n_0}(t) = e^{-\lambda n_0 t}.$$

Armed with this knowledge of p_{n_0} we can now find $p_{n_0 + 1}$. The differential equation (9.8) is, with $n = n_0 + 1$,

$$\frac{dp_{n_0 + 1}}{dt} + \lambda(n_0 + 1)p_{n_0 + 1} = \lambda n_0 p_{n_0}. \qquad (9.10)$$

This will be recognized as a linear first-order differential equation in standard form (see any first-year calculus text). Its integrating factor is $e^{\lambda(n_0 + 1)t}$ and in Exercise 10 it is shown that

$$p_{n_0 + 1}(t) = n_0 e^{-\lambda n_0 t}(1 - e^{-\lambda t}). \qquad (9.11)$$

Having obtained $p_{n_0 + 1}$ we can solve the equation for $p_{n_0 + 2}$, etc. In general we obtain for the probability that there have been a total number of k births at t,

$$\boxed{p_{n_0 + k}(t) = \binom{n_0 + k - 1}{n_0 - 1} e^{-\lambda n_0 t}(1 - e^{-\lambda t})^k} , \qquad k = 0, 1, 2, \ldots, \qquad (9.12)$$

as will be verified in Exercise 11.

It can be seen that $p_{n_0}(t)$ is an exponentially decaying function of time. If we define T_1 as the **time of the first birth** and observe that $p_{n_0}(t)$ is the probability of no birth in $(0, t]$ we get

$$\Pr\{T_1 > t\} = e^{-\lambda n_0 t},$$

or equivalently

$$\Pr\{T_1 \leqslant t\} = 1 - e^{-\lambda n_0 t}.$$

Hence T_1 is exponentially distributed and has mean

$$E(T_1) = \frac{1}{\lambda n_0}.$$

We notice that the larger n_0 is, the faster does $p_{n_0}(t)$ decay towards zero – as it must because the larger the population, the greater the chance for a birth. Note also that $p_{n_0}(t)$ is never zero for $t < \infty$. Thus there is always a non-zero probability that the population will remain unchanged in any finite time interval.

A plot of $p_{n_0}(t)$ versus t when $\lambda = n_0 = 1$ is shown in Fig. 9.3. Also shown are the graphs of $p_{n_0 + 1}(t)$ and $p_{n_0 + 2}(t)$ which rise from zero to achieve maxima before declining to zero at $t = \infty$. In Exercise 12 it is shown that for these parameter values $p_{1 + k}(t)$ has a maximum at $t = \ln(1 + k)$.

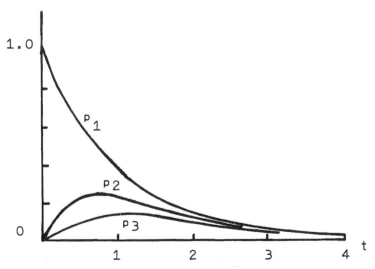

Figure 9.3 Probabilities of no births, one birth and two births as functions of time in the Yule process with $\lambda = 1$ and initially $n_0 = 1$ individual.

9.5 MEAN AND VARIANCE FOR THE YULE PROCESS

For the simple birth process described in the previous section we will show that the mean population size $\mu(t) = E(N(t)| N(0) = n_0)$ at time t is

$$\mu(t) = n_0 e^{\lambda t}.$$

This is the same as the Malthusian growth law with no deaths and a birth rate $b = \lambda$.

Proof By definition

$$\mu(t) = \sum_{n = n_0}^{\infty} np_n(t).$$

Differentiate to get

$$\frac{d\mu}{dt} = \sum_{n = n_0}^{\infty} n\frac{dp_n}{dt} = n_0\frac{dp_{n_0}}{dt} + \sum_{n = n_0 + 1}^{\infty} n\frac{dp_n}{dt}.$$

Substituting from the differential and differential-difference equations (9.9) and (9.8),

$$\frac{d\mu}{dt} = - \lambda n_0^2 p_{n_0} + \lambda \sum_{n = n_0 + 1}^{\infty} [(n^2 - n)p_{n - 1} - n^2 p_n].$$

The coefficient of p_{n-1} is now rewritten to contain a perfect square:

$$\frac{d\mu}{dt} = - \lambda n_0^2 p_{n_0} + \lambda \sum_{n = n_0 + 1}^{\infty} (n - 1)^2 p_{n - 1}$$

$$+ \lambda \sum_{n = n_0 + 1}^{\infty} (n - 1)p_{n - 1} - \lambda \sum_{n = n_0 + 1}^{\infty} n^2 p_n.$$

Put $m = n - 1$ in the first two sums on the right and get

$$\frac{d\mu}{dt} = \left(- \lambda n_0^2 p_{n_0} + \lambda \sum_{m = n_0}^{\infty} m^2 p_m - \lambda \sum_{n = n_0 + 1}^{\infty} n^2 p_n \right) + \lambda \sum_{m = n_0}^{\infty} mp_m.$$

The terms in brackets cancel to leave the first-order differential equation

$$\frac{d\mu}{dt} = \lambda \sum_{m = n_0}^{\infty} mp_m = \lambda\mu, \tag{9.13}$$

which has the initial condition

$$\mu(0) = n_0. \tag{9.14}$$

Integrating (9.13) and using (9.14) gives the required result. Similarly, the second moment of $N(t)$ may be found and hence the variance. However, it is quicker to use the properties of the negative binomial distribution.

Mean and variance from the negative binomial distribution

Consider a sequence of Bernoulli trials with probability p of success and probability $q = 1 - p$ of failure at each trial. Let the random variable X_r be the number of trials up to and including the rth success, $r = 1, 2, \ldots$. Then it is easily seen that the distribution of X_r is given by

$$\Pr\{X_r = k\} = \binom{k-1}{r-1} p^r q^{k-r}, \qquad k = r, r+1, r+2, \ldots \qquad (9.15)$$

the smallest possible value of X_r being r since there must be at least r trials to obtain r successes. The mean and variance of X_r are found to be (see Exercise 13),

$$E(X_r) = \frac{r}{p},$$

$$\mathrm{Var}(X_r) = \frac{rq}{p^2}.$$

We now put $j = n_0 + k$ in (9.12) to get

$$\Pr\{N(t) = j\} = \binom{j-1}{n_0-1} e^{-\lambda n_0 t}(1 - e^{-\lambda t})^{j-n_0}, \quad j = n_0, n_0 + 1, n_0 + 2, \ldots$$

This is seen to be a negative binomial distribution as in (9.15) with parameters

$$r = n_0,$$
$$p = e^{-\lambda t}, q = 1 - e^{-\lambda t}.$$

Thus we quickly see that the mean and variance of the population in the Yule process are

$$\boxed{\begin{aligned} E(N(t)) &= n_0 e^{\lambda t} \\ \mathrm{Var}(N(t)) &= n_0(1 - e^{-\lambda t})e^{2\lambda t} \end{aligned}}$$

Approximations

(i) *Large t* For large t the variance is asymptotically

$$\mathrm{Var}(N(t)) \underset{t \to \infty}{\sim} n_0 e^{2\lambda t},$$

and the standard deviation of the population is

$$\sigma(t) \sim \sqrt{n_0}\, e^{\lambda t}.$$

Thus the mean and standard deviation grow, as indicated in Fig. 9.4, and eventually their ratio is constant.

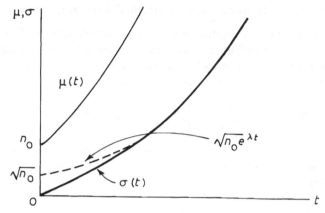

Figure 9.4 Mean and standard deviation of the population as functions of time in a Yule process.

(ii) *Large n_0* Let us suppose that there is just one individual to start with. The probability that there is still only one individual at time t, an event we will call a 'failure', is

$$\Pr\{\text{population is unchanged at } t\} = e^{-\lambda t} = \Pr\{\text{'failure'}\}.$$

Now, if there are $n_0 > 1$ individuals to start with, since individuals act independently,

$$\Pr\{\text{population is unchanged at } t\} = \Pr\{n_0 \text{ failures in } n_0 \text{ trials}\}$$
$$= (e^{-\lambda t})^{n_0} = e^{-\lambda n_0 t},$$

this being an exact result. Also, when $n_0 = 1$ let the mean and variance of the population be denoted by μ_1 and σ_1^2 which are given by

$$\mu_1 = e^{\lambda t}, \sigma_1^2 = e^{2\lambda t}(1 - e^{-\lambda t}).$$

When $n_0 > 1$ we may at any time divide the population into n_0 groups, those in each group being descendants of one of the original individuals. The number in each group is a random variable and it follows that the population at time t is the sum of n_0 independent and identically distributed random variables each having mean μ_1 and variance σ_1^2. By the central limit theorem we see that for large n_0 and any t, the random variable $N(t)$ is approximately normal with mean $n_0\mu_1$ and variance $n_0\sigma_1^2$. Hence we may estimate with reasonable accuracy the probability that the population lies within prescribed limits (see exercises).

9.6 A SIMPLE DEATH PROCESS

In this rather macabre continuous time Markov chain, individuals persist only until they die and there are no replacements. The assumptions are similar to those in the birth process of the previous two sections, but now each individual, if still alive at time t, is removed in $(t, t + \Delta t]$ with probability $\mu \Delta t + o(\Delta t)$. Again we are interested in finding the transition probabilities

$$p_n(t) = \Pr\{N(t) = n \mid N(0) = n_0\}, \qquad n = n_0, n_0 - 1, \ldots, 2, 1, 0.$$

We could proceed via differential-difference equations for p_n but there is a more expeditious method.

The case of one individual

Let us assume $n_0 = 1$. Now $p_1(t)$ is the probability that this single individual is still alive at t and we see that

$$p_1(t + \Delta t) = p_1(t)(1 - \mu \Delta t) + o(\Delta t) \tag{9.16}$$

since $1 - \mu \Delta t$ is the probability that the individual did not die in $(t, t + \Delta t]$. From (9.16) it quickly follows that

$$\frac{dp_1}{dt} = -\mu p_1, \qquad t > 0.$$

The solution with initial value $p_1(0) = 1$ is just

$$\boxed{p_1(t) = e^{-\mu t}}$$

The initial population size is $N(0) = n_0 > 1$

If there are n_0 individuals at $t = 0$, the number alive at t is a binomial random variable with parameters n_0 and $p_1(t)$. Therefore we have immediately

$$\boxed{p_n(t) = \binom{n_0}{n} e^{-\mu n t}(1 - e^{-\mu t})^{n_0 - n}}, \qquad n = n_0, n_0 - 1, \ldots, 1, 0.$$

Also

$$E(N(t)) = n_0 e^{-\mu t},$$

which corresponds to a Malthusian growth law with $d = \mu$ and $b = 0$, and

$$\mathrm{Var}\,(N(t)) = n_0 e^{-\mu t}(1 - e^{-\mu t}).$$

Extinction

In this pure death process the population either remains constant or it decreases. It may eventually reach zero in which case we say that the population has gone *extinct*. The probability that the population is extinct at time t is

$$\Pr\{N(t) = 0 \mid N(0) = n_0\} = (1 - e^{-\mu t})^{n_0} \to 1 \quad \text{as} \quad t \to \infty.$$

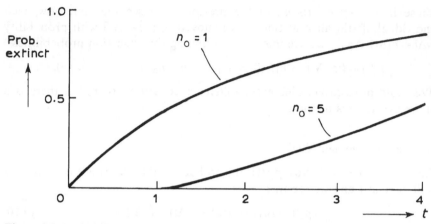

Figure 9.5 Probabilities that the population is extinct at t, versus t for various initial populations with $\mu = 0.5$.

Thus extinction is inevitable in this model. In Fig. 9.5 are shown the probabilities of extinction versus time for various initial populations.

9.7 SIMPLE BIRTH AND DEATH PROCESS

We now combine the ideas of the Yule process and the simple death process of the previous section. Let there be n_0 individuals initially and $N(t)$ at time t. In $(t, t + \Delta t]$ an individual has an offspring with probability $\lambda \Delta t + o(\Delta t)$ and dies with probability $\mu \Delta t + o(\Delta t)$. Using the same kind of reasoning as in Section 9.4 for the population birth probabilities we find that

$$\Pr\{\text{one birth in } (t, t + \Delta t] \mid N(t) = n\} = \lambda n \Delta t + o(\Delta t)$$
$$\Pr\{\text{one death in } (t, t + \Delta t] \mid N(t) = n\} = \mu n \Delta t + o(\Delta t)$$
$$\Pr\{\text{no change in population size in } (t, t + \Delta t] \mid N(t) = n\}$$
$$= 1 - (\lambda + \mu)n \Delta t + o(\Delta t).$$

The ways to obtain a population size n at time $t + \Delta t$ are, if $n \geqslant 1$,

$$\begin{cases} N(t) = n - 1 \text{ and one birth in } (t, t + \Delta t] \\ N(t) = n \text{ and no change in } (t, t + \Delta t] \\ N(t) = n + 1 \text{ and one death in } (t, t + \Delta t]. \end{cases}$$

Hence

$$p_n(t + \Delta t) = p_{n-1}(t)\lambda(n-1)\Delta t + p_n(t)[1 - (\lambda + \mu)n\Delta t]$$
$$+ p_{n+1}(t)\mu(n+1)\Delta t + o(\Delta t).$$

It quickly follows that

$$\boxed{\frac{dp_n}{dt} = \lambda(n-1)p_{n-1} - (\lambda + \mu)np_n + \mu(n+1)p_{n+1}}, \qquad n \geqslant 1. \qquad (9.17)$$

If $n = 0$ we have simply

$$\frac{dp_0}{dt} = \mu p_1, \qquad (9.18)$$

and the initial conditions are

$$p_n(0) = \begin{cases} 1, & n = n_0, \\ 0, & n \neq n_0. \end{cases}$$

The system of equations (9.17) and (9.18) cannot be solved recursively as could the equations for the simple birth (Yule) process as there is no place to get started.

The probability generating function of $N(t)$

By definition, the probability generating function of $N(t)$ is

$$\phi(s, t) = \sum_{n=0}^{\infty} p_n(t)s^n.$$

This can be shown (see Exercise 16) to satisfy the first-order partial differential equation

$$\frac{\partial \phi}{\partial t} = (\lambda s - \mu)(s - 1)\frac{\partial \phi}{\partial s}. \qquad (9.19)$$

which is to be solved with the initial condition

$$\phi(s, 0) = s^{n_0}. \qquad (9.20)$$

It may be shown (see, for example, Pollard, 1973; Bailey, 1964) and it will be verified in Exercise 17, that the solution of (9.19) and (9.20) is

$$\phi(s, t) = \left(\frac{\mu - \psi(s)e^{-(\lambda - \mu)t}}{\lambda - \psi(s)e^{-(\lambda - \mu)t}}\right)^{n_0}, \qquad (9.21)$$

where

$$\psi(s) = \frac{\lambda s - \mu}{s - 1}. \qquad (9.22)$$

The probability of extinction

A few sample paths of the simple birth and death process are shown in Fig. 9.6. The state space is the set of all non-negative integers $\{0, 1, 2, \ldots\}$ and the state 0

Figure 9.6 Two representative sample paths of a birth and death process. Here $N(0) = 6$ and one path is shown on which the population goes extinct.

is clearly absorbing. A sample path may terminate at 0 which corresponds to extinction of the population.

We can easily find the probability that extinction has occurred at or before time t from the probability generating function. This is just

$$\Pr\{N(t) = 0 \mid N(0) = n_0\} = p_0(t) = \phi(0, t).$$

From (9.22) we have $\psi(0) = \mu$ and thence, from (9.21),

$$\phi(0, t) = \left(\frac{\mu(1 - e^{-(\lambda - \mu)t})}{\lambda - \mu e^{-(\lambda - \mu)t}}\right)^{n_0}, \qquad \lambda \neq \mu. \tag{9.23}$$

When $\lambda = \mu$ the following expression is obtained by taking the appropriate limit in (9.23):

$$\phi(0, t) = \left(\frac{\lambda t}{\lambda t + 1}\right)^{n_0}, \qquad \lambda = \mu. \tag{9.24}$$

In the limit $t \to \infty$, $\phi(0, t)$ approaches the probability that the population ever goes extinct. Denote this quantity by p_{ext}. Then from (9.23) and (9.24) we find

$$p_{ext} = \begin{cases} 1, & \lambda \leqslant \mu \\ \left(\dfrac{\mu}{\lambda}\right)^{n_0}, & \lambda > \mu \end{cases}$$

Thus extinction is inevitable if the probability of a birth to any individual is less than or equal to the probability of death in any small time interval. It may seem surprising that extinction is certain when $\lambda = \mu$. To understand this we note that 0 is an absorbing barrier which is always a finite distance from the value of $N(t)$. The situation is similar in the random walk on $[0, 1, 2, \ldots)$ with $p = q$ where we found that absorption at 0 is certain (see Section 7.6).

In the cases $\lambda \leqslant \mu$ where extinction is certain, we may define the random variable T which is the **extinction time**. Evidently the distribution function of T is

$$\Pr\{T \leqslant t\} = \phi(0, t)$$

since this is the probability that extinction occurs at or before t. When $\lambda = \mu$ the expected extinction time is infinite (see Exercise 18) but it is finite when $\lambda < \mu$. When $\lambda > \mu$ we may still talk of the random variable T, the extinction time. However, we then have

$$\Pr\{T < \infty\} = \left(\frac{\mu}{\lambda}\right)^{n_0},$$

so we must also have

$$\Pr\{T = \infty\} = 1 - \left(\frac{\mu}{\lambda}\right)^{n_0}.$$

Clearly in these cases T has no finite moments and, because its probability mass is not all concentrated on $(0, \infty)$ we say it is not a **'proper' random variable**.

9.8 MEAN AND VARIANCE FOR THE BIRTH AND DEATH PROCESS

The expected number of individuals at time t is

$$m(t) = E[N(t) | N(0) = n_0] = \sum_{n=0}^{\infty} n p_n(t) = \sum_{n=1}^{\infty} n p_n(t).$$

We will find a differential equation for $m(t)$. We have

$$\frac{dm}{dt} = \sum_{n=1}^{\infty} n \frac{dp_n}{dt}$$

and on substituting from the differential-difference equation (9.17) we get

$$\frac{dm}{dt} = \sum_{n=1}^{\infty} n[\lambda(n-1)p_{n-1} - (\lambda + \mu)np_n + \mu(n+1)p_{n+1}]$$

which rearranges to

$$\frac{dm}{dt} = \lambda \sum_{n=1}^{\infty} (n-1)^2 p_{n-1} + \lambda \sum_{n=1}^{\infty} (n-1)p_{n-1} - (\lambda + \mu) \sum_{n=1}^{\infty} n^2 p_n$$

$$+ \mu \sum_{n=1}^{\infty} (n+1)^2 p_{n+1} - \mu \sum_{n=1}^{\infty} (n+1)p_{n+1}.$$

A relabeling of indices with $n' = n - 1$ in sums involving p_{n-1} and with $n'' = n + 1$ in sums involving p_{n+1} yields

$$\frac{dm}{dt} = -(\lambda + \mu) \sum_{n=1}^{\infty} n^2 p_n + \lambda \sum_{n=0}^{\infty} n'^2 p_{n'} + \mu \sum_{n''=2}^{\infty} n''^2 p_{n''}$$

$$+ \lambda \sum_{n'=0}^{\infty} n' p_{n'} - \mu \sum_{n''=2}^{\infty} n'' p_n$$

In the first three sums here, terms from $n, n', n'' = 2$ and onward cancel and leave $-(\lambda + \mu)p_1 + \lambda p_1 = -\mu p_1$. Thus

$$\frac{dm}{dt} = -\mu p_1 + \lambda \sum_{n'=0}^{\infty} n' p_{n'} - \mu \sum_{n'=2}^{\infty} n'' p_{n''}$$

$$= (\lambda - \mu) \sum_{n=0}^{\infty} n p_n$$

or simply

$$\frac{dm}{dt} = (\lambda - \mu)m.$$

With initial condition $m(0) = n_0$ the solution is

$$\boxed{m(t) = n_0 e^{(\lambda - \mu)t}.}$$

This is the same as the deterministic result (Malthusian law) of Section 9.1 with the birth rate b replaced by λ and the death rate d replaced by μ.

The second moment of $N(t)$,

$$M(t) = \sum_{n=0}^{\infty} n^2 p_n(t)$$

can be shown to satisfy

$$\frac{dM}{dt} = 2(\lambda - \mu)M + (\lambda + \mu)M, \quad M(0) = n_0^2, \tag{9.25}$$

as will be seen in Exercise 19. The variance of the population in the birth and death process may then be shown to be

$$\boxed{\mathrm{Var}\,(N(t)|\,N(0) = n_0) = n_0 \frac{(\lambda + \mu)}{(\lambda - \mu)} e^{(\lambda - \mu)t} [e^{(\lambda - \mu)t} - 1]}, \quad \lambda \neq \mu.$$

In the special case $\lambda = \mu$,

$$\boxed{\mathrm{Var}\,(N(t)|\,N(0) = n_0) = 2\lambda n_0 t.}$$

An alternative method of finding the moments of $N(t)$ is to use the moment generating function (see Exercise 20).

Birth and death processes have recently become very important in studies of how ions move across cell membranes. In the simplest model there are just two states for an **ion channel** – open and closed. The channel stays in each state for an exponentially distributed time before making a transition to the other state. It is hoped that a study of such continuous time Markov chain models will elucidate the mechanisms by which molecules of the membrane interact with various drugs. For details of this fascinating application see Colquhoun and Hawkes (1977), Hille (1984) and Tuckwell (1988).

REFERENCES

Bailey, N.T.J. (1964). *The Elements of Stochastic Processes*. Wiley, New York.

Bartlett, M. S. (1960). *Stochastic Population Models*. Methuen, London.

Cameron, R. J. (ed.) (1982). *Year Book Australia*. Australian Bureau of Statistics, Canberra.

Colquhoun, D. and Hawkes, A.G. (1977). Relaxations and fluctuations of membrane currents that flow through drug operated channels. *Proc. R. Soc. Lond. B.*, **199**, 231-62.

Furry, W.H. (1937). On fluctuation phenomena in the passage of high-energy electrons through lead. *Phys. Rev.*, **52**, 569-81.

Hille, B. (1984). *Ionic Channels of Excitable Membranes*. Sinauer, Sunderland, Mass.

Keyfitz, N. (1968). *Introduction to the Mathematics of Populations*. Addison-Wesley, Reading, Mass.

Ludwig, D. (1978). *Stochastic Population Theories*. Springer, New York.

Parzen, E. (1962). *Stochastic Processes*. Holden-Day, San Francisco.

Pielou, E. C. (1969). *An Introduction to Mathematical Ecology*. Wiley, New York.

Pollard, J.H. (1973). *Mathematical Models for the Growth of Human Populations*. Cambridge University Press, London.

Tuckwell, H.C. (1981). Poisson processes in biology. In *Stochastic Nonlinear Systems*, Springer-Verlag, Berlin, pp. 162-71.

Tuckwell, H.C. (1988). *Stochastic Processes in the Neurosciences*. SIAM, Philadelphia.

Yule, G.U. (1924). A mathematical theory of evolution based upon the conclusions of Dr J.C. Willis, F.R.S. *Phil. Trans. Roy. Soc. Lond. B.*, **213**, 21-87.

EXERCISES

1. Using the birth and death rates for 1966 given in Table 9.2 and the 1966 population of Australia given in Table 9.1, estimate the 1971 population. Compare with the actual population in 1971. Is the discrepancy in the direction you would expect? Why?

2. In a simple Poisson process, let $p_n(t) = \Pr\{N(t) = n \mid N(0) = 0\}$. Use the relations

$$\Pr\{\Delta N(t) = k\} = \begin{cases} 1 - \lambda\Delta t + o(\Delta t), & k = 0, \\ \lambda\Delta t + o(\Delta t), & k = 1, \\ o(\Delta t), & k \geqslant 2, \end{cases}$$

to derive differential-difference equations for p_n, $n = 0.1, 2, \ldots$, in the same manner in which (9.8) was derived for the Yule process. Solve the system of equations to recover the Poisson probabilities (9.1).

3. Show how the defining properties of a simple Poisson process enable the joint distribution of $\{N(t_1), N(t_2), \ldots, N(t_n)\}$ to be found for arbitrary $0 \leqslant t_1 < t_2 < \cdots < t_n < \infty$.

4. What, if any, are the differences between a simple Poisson process and a Poisson point process?

5. Name as many as you can of the deficiencies of the simple Poisson process as a realistic model for the growth of a population of, say, humans.

6. If $\{N(t), t \geqslant 0\}$ is a simple Poisson process, find the characteristic function of $N(t)$.

7. Let N_1 and N_2 be two independent simple Poisson processes with rate parameters λ_1 and λ_2 respectively. Define a new process $X = \{X(t), t \geqslant 0\}$ by

$$X(t) = N_1(t) + N_2(t).$$

(i) Find $E(X(t))$ and $\mathrm{Var}(X(t))$.
(ii) Is X a Poisson process?

8. For a continuous time Markov chain the transition probabilities are (equation (9.3)),

$$p(s_n, t_n | s_{n-1}, t_{n-1}) = \Pr\{X(t_n) = s_n | X(t_{n-1}) = s_{n-1}\}.$$

Show that a Poisson process is a continuous time Markov chain but that the only transition probabilities needed are $p(k, t | 0, 0)$, $k = 0, 1, 2, \ldots, t > 0$.

9. Let X be a continuous time Markov chain with stationary transition probabilities

$$\Pr\{X(t) = s_k | X(0) = s_j\} \doteq p(k, t | j).$$

Give an interpretation of the Chapman–Kolmogorov relation

$$p(k, t_1 + t_2 | i) = \sum_j p(j, t_1 | i) p(k, t_2 | j), \quad t_1, t_2 > 0$$

using a path diagram.

10. In the Yule process, $p_{n_0 + 1}(t)$ is the probability that the population has increased by one at time t. This quantity satisfies (9.10); i.e.,

$$p'_{n_0 + 1} + \lambda(n_0 + 1)p_{n_0 + 1} = \lambda n_0 p_{n_0}, \quad p_{n_0 + 1}(0) = 0.$$

Show that the solution of this equation is as given by (9.11).

11. For the Yule process, prove that $p_{n_0 + k}(t)$, $k = 0, 1, 2, \ldots$ is given by (9.12).

12. In a simple birth process in which $\lambda = n_0 = 1$, show that $p_{1 + k}(t)$ has a maximum at $t = \ln(1 + k)$.

13. Prove that a negative binomial random variable X_r, with probability law

given by (9.15), has mean and variance

$$E(X_r) = \frac{r}{p}, \qquad \text{Var}(X_r) = \frac{rq}{p^2}.$$

(*Hint*: X_r is the sum of r i.i.d. random variables.)

14. A herd of 50 ungulates is released on to a large island. The birth and death probabilities are $\lambda = 0.15$ and $\mu = .05$ per animal per year. A hunter wishes to visit the island when he can be 95% sure that the herd has doubled in size. How long will he have to wait? Assume a simple birth and death process applies.

15. Let T be the extinction time in a pure death process with n_0 individuals initially. What is the density of T?

16. Show that the probability generating function $\phi(s,t)$ for $\Pr\{N(t) = n \mid N(0) = n_0\}$ in the simple birth and death process satisfies

$$\frac{\partial \phi}{\partial t} = (\lambda s - \mu)(s - 1)\frac{\partial \phi}{\partial s}.$$

17. Verify, by direct substitution, that the function $\phi(s,t)$ given in (9.21) satisfies the partial differential equation of Exercise 16 with initial data $\phi(s,0) = s^{n_0}$.

18. Show that when $\lambda = \mu$ in the simple birth and death process, the expectation of the extinction time is infinite.

19. If $M(t)$ is the second moment, $E[N^2(t) \mid N(0) = n_0]$ in the simple birth and death process, prove using the differential-difference equations (9.17), that M satisfies (9.25).

20. Let the moment generating function of $N(t)$ in the birth and death process be $\psi(\theta, t) = E(e^{\theta N(t)}) = \phi(e^{\theta}, t)$. From the given expression for $\phi(s,t)$ in (9.21), find $\psi(\theta, t)$ and hence the mean and variance of $N(t)$.

10

Population growth II: branching processes

10.1 CELL DIVISION

In Section 2.3 we mentioned the phenomenon of mitosis whereby two cells emerge from one. The underlying complex sequence of biochemical reactions is orchestrated by the molecules of heredity known as DNA (deoxyribonucleic acid). It is truly ponderable that each of the 10^{14} cells of the average adult human are all, without a single exception, derived by mitosis from the single cell which results from the combination of female and male gametes. The descendants from a single cell are referred to as a **clone**. Some reproducing cells are shown in Fig. 10.1.

Mitotic events continue throughout the life of the adult and are in fact essential for it. For example, red blood cells have an average lifetime of 120 days. There are about 2.5×10^{13} such cells in the 5 litres of adult blood and these have to be continually replaced at the rate of 2.5×10^6 per second. Altogether there are about 2×10^7 cell divisions per second in the adult human (Prescott, 1976).

On the tragic side, mitosis in adults or even children may proceed at an abnormally high rate to produce cancer. Thus one abnormal cell may undergo division to form two abnormal cells which in turn inherit the property of faster than normal mitosis. This may give rise to a clone of billions of cancerous cells which may lead to the death of the host. The most alarming aspect of this is that it only requires the presence of one abnormal cell out of the many billions in the body which are capable of mitosis. Such abnormal cells may be produced by the action on normal cells of **carcinogens** such as radiation, cigarette smoke, asbestos, pesticides, etc. See Kimball (1984) for an interesting overview.

In a population of cells the time at which mitosis occurs in any cell is unpredictable. In fact, according to analyses by cell biologists, a truly random event or sequence of events may precede mitosis (Brooks, 1981). However, the time it takes a cell to divide may also correlate with size as Fig. 10.2 illustrates.

Figure 10.1 Reproducing yeast cells. Cells which are about to give rise to two cells are in the form of a figure-of-eight. From Kimball (1984).

10.2 THE GALTON - WATSON BRANCHING PROCESS

The process of formation of a clone of cells has an analogy in the inheritance of family names. Indeed, this was the area in which **branching processes** were first considered mathematically. In 1873, Galton posed the following problem in the *Educational Times*:

Problem 4001: A large nation, of whom we will only concern ourselves with the adult males, N in number, and who each bear separate surnames, colonise a district. Their law of population is such that, in each generation, a_0 per cent of the adult males have no male children who reach adult life; a_1 have one such male child; a_2 have two; and so on up to a_5 who have five.
Find (1) what proportion of the surnames will have become extinct after r

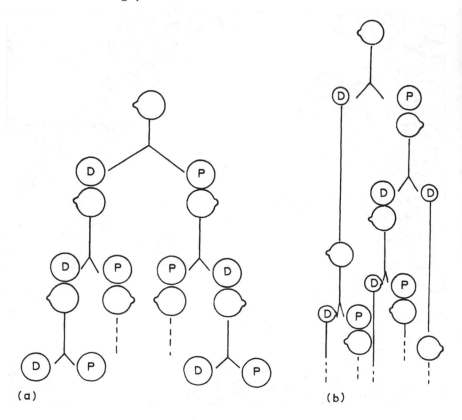

Figure 10.2 Diagrammatic representation of two clones of yeast cells. Time increases downwards. In (a) parent and daughter cells are about the same size and form buds almost simultaneously. In (b) daughter cells are smaller and must grow before a bud appears. From Carter (1981).

generations; and (2) how many instances there will be of the same surname being held by m persons.

The problem was formulated because it had been noticed that the majority of family names, and in particular those of the nobility, tended to die out after a few hundred years. It was desirable to ascertain whether this was a random phenomenon or was due to the 'diminution in fertility'. The problem was partially solved by Watson in 1874 (Watson and Galton, 1874) but not completely until 1930 by Steffenson.

Definition Let $n = 0, 1, 2, \ldots$ be the generation number. A Galton–Watson branching process is a random process $X = \{X_n, n = 0, 1, 2, \ldots\}$ such that:

(i) There is initially one individual so $X_0 = 1$.

(ii) The numbers of immediate descendants of the individuals in any generation are identically distributed independent random variables taking on values $0, 1, 2, \ldots$ with probabilities $p_0, p_1, p_2, \ldots$.

It is clear that X is a Markov process because if the value of X_k is known for some k, we can determine the probabilities that $X_{k+1}, X_{k+2}, \ldots$ take on various values without any knowledge of the population sizes at generations before the kth.

To describe the relation between X_{n+1} and X_n we introduce an infinite matrix of independent and identically distributed random variables $\{Z_{ij}, i = 1, 2, \ldots; j = 0, 1, 2, \ldots\}$, where Z_{ij} is the number of descendants of the ith individual in the jth generation. For $j = 0$ only Z_{10} is needed, this being the number of descendants of the individual present at the beginning. The distributions of the Z_{ij} are given by

$$\Pr\{Z_{10} = k\} = p_k, \qquad k = 0, 1, 2, \ldots$$

and we put

$$E(Z_{10}) = \mu$$
$$\operatorname{Var}(Z_{10}) = \sigma^2.$$

The evolution of the branching process is then described by the recursive system

$$X_{n+1} = \sum_{k=1}^{X_n} Z_{kn}, \qquad n = 0, 1, 2, \ldots$$

For $n = 0$ there is one term in the sum, namely Z_{10}. For $n \geqslant 1$ there are a random number of terms in the sum, each term being a random variable (cf. Section 3.7). If perchance $X_n = 0$ for some n, the sum is interpreted as zero.

10.3 MEAN AND VARIANCE FOR THE GALTON-WATSON PROCESS

To find the mean and variance of the number of individuals, X_n, comprising the nth generation, we proceed in the same fashion as in Section 3.3. Here, however, we first seek a recursive relation between the first two moments of two consecutive generations.

We have

$$E(X_{n+1}) = E\left(\sum_{k=1}^{X_n} Z_{kn}\right)$$
$$= \sum_j E\left(\sum_{k=1}^{X_n} Z_{kn} \,\middle|\, X_n = j\right) \Pr\{X_n = j\},$$

by the law of total probability applied to expectations. Now, if $X_n = j$ the population X_{n+1} is the sum of j random variables each with mean μ. Hence

$$E(X_{n+1}) = \sum_j j\mu \Pr\{X_n = j\}$$

so

$$E(X_{n+1}) = \mu E(X_n), \qquad n = 0, 1, 2, \ldots$$

Since $E(X_0) = 1$ we find

$$E(X_1) = \mu$$

$$E(X_2) = \mu E(X_1) = \mu^2$$

$$\boxed{E(X_n) = \mu^n}, \qquad n = 0, 1, 2, \ldots$$

To find a relation between $E(X_{n+1}^2)$ and $E(X_n^2)$ we note firstly that

$$\mathrm{Var}(X_{n+1} \mid X_n = j) = \mathrm{Var}\left(\sum_{k=1}^{X_n} Z_{nk} \mid X_n = j\right) = j\sigma^2,$$

because the Z_{nk} are independent. Hence

$$E(X_{n+1}^2 \mid X_n = j) = E\left(\left(\sum_{k=1}^{X_n} Z_{kn}\right)^2 \mid X_n = j\right) = j\sigma^2 + j^2\mu^2.$$

Substituting this in

$$E(X_{n+1}^2) = \sum_j E(X_{n+1}^2 \mid X_n = j) \Pr\{X_n = j\}$$

gives

$$E(X_{n+1}^2) = \sigma^2 \sum_j j \Pr\{X_n = j\} + \mu^2 \sum_j j^2 \Pr\{X_n = j\}$$

$$= \sigma^2 E(X_n) + \mu^2 E(X_n^2).$$

We now arrive at

$$\mathrm{Var}(X_{n+1}) + E^2(X_{n+1}) = \sigma^2\mu^n + \mu^2[\mathrm{Var}(X_n) + E^2(X_n)]$$

which reduces to the following recursion relation for the variance:

$$\mathrm{Var}(X_{n+1}) = \sigma^2\mu^n + \mu^2\,\mathrm{Var}(X_n), \qquad n = 0, 1, 2, \ldots \qquad (10.1)$$

In Exercise 1 it is deduced from this that

$$\boxed{\mathrm{Var}(X_n) = \begin{cases} \dfrac{\sigma^2\mu^{n-1}(\mu^n - 1)}{\mu - 1}, & \mu \neq 1 \\[2mm] n\sigma^2, & \mu = 1 \end{cases}} \qquad n = 0, 1, 2, \ldots$$

The mean and variance have the following interesting asymptotic behaviour as $n \to \infty$.

(i) $\mu < 1$

$$E(X_n) \to 0,$$
$$\text{Var}(X_n) \to 0.$$

(ii) $\mu = 1$

$$E(X_n) = 1 \quad \text{for all} \quad n,$$
$$\text{Var}(X_n) \to \infty.$$

(iii) $\mu > 1$

$$E(X_n) \to \infty,$$
$$\text{Var}(X_n) \to \infty.$$

Note in particular that in case (ii), when the expected number of replacements of an individual is one, the expected population size remains finite although the variance becomes infinite.

Before attacking the next problem of interest we need the following useful digression.

10.4 PROBABILITY GENERATING FUNCTIONS OF SUMS OF RANDOM VARIABLES

The probability that a Galton - Watson process goes extinct will be found from a relation between the **probability generating functions** (p.g.f.s) of X_n and X_{n-1}. To establish this we first need some general preliminary results on the p.g.f.s of sums of random variables. We begin with a theorem involving just two non-negative integer-valued random variables X and Y whose p.g.f.s are, for suitable s,

$$f(s) = \sum_{k=0}^{\infty} \Pr\{X = k\}s^k \doteq \sum_{k=0}^{\infty} f_k s^k \tag{10.2}$$

$$g(s) = \sum_{k=0}^{\infty} \Pr\{Y = k\}s^k \doteq \sum_{k=0}^{\infty} g_k s^k \tag{10.3}$$

Theorem 10.1 If X and Y are independent random variables with p.g.f.s given by (10.2) and (10.3), then their sum

$$Z = X + Y$$

has p.g.f.

$$h(s) = f(s)g(s)$$

Thus, the p.g.f. of the sum of the two independent random variables is the product of their p.g.f.s.

Proof By definition the coefficient of s^k in $h(s)$ is h_k. That is,

$$h(s) = \sum_{k=0}^{\infty} \Pr\{Z = k\}s^k \doteq \sum_{k=0}^{\infty} h_k s^k.$$

We need to show the coefficient of s^k in $f(s)g(s)$ is also h_k. We have

$$f(s)g(s) = \left(\sum_{i=0}^{\infty} f_i s^i \right) \left(\sum_{j=0}^{\infty} g_j s^j \right),$$

and the coefficient of s^k is

$$f_k g_0 + f_{k-1} g_1 + \cdots + f_0 g_k. \tag{10.4}$$

Also

$$h_k = \Pr\{Z = k\} = \sum_{i=0}^{k} \Pr\{Z = k \mid Y = i\} \Pr\{Y = i\}$$

$$= \sum_{i=0}^{k} \Pr\{X = k - i\} \Pr\{Y = i\}$$

$$= \sum_{i=0}^{k} f_{k-i} g_i.$$

This sum, called **convolution**, is the same as (10.4) so the theorem is proved.

The following result is an extension of Theorem 10.1 to the case of a sum of more than two random variables.

Corollary Let X_1, X_2,..., X_n be i.i.d. non-negative integer-valued random variables with common generating function $f(s)$. Then their sum

$$S_n = X_1 + X_2 + \cdots + X_n \tag{10.5}$$

has p.g.f.

$$h(s) = f^n(s).$$

The sum in (10.5) has a definite number of terms. We now turn to the case of a sum whose number of terms is random, as we encountered in Section 3.3.

Theorem 10.2 Let $\{X_k, k = 1, 2, ...\}$ be i.i.d. non-negative integer-valued random variables with common p.g.f.

$$f(s) = \sum_{i=0}^{\infty} \Pr\{X_1 = i\}s^i \doteq \sum_{i=0}^{\infty} f_i s^i,$$

and let N be a non-negative integer-valued random variable, independent of the X_k, and with p.g.f.

$$g(s) = \sum_{n=0}^{\infty} \Pr\{N=n\}s^n \doteq \sum_{n=0}^{\infty} g_n s^n.$$

Define the random sum

$$S_N = X_1 + X_2 + \cdots + X_N$$

with the agreement that $S_N = 0$ if $N = 0$. Then the p.g.f. of S_N,

$$h(s) = \sum_{j=0}^{\infty} \{S_N = j\}s^j \doteq \sum_{j=0}^{\infty} h_j s^j$$

is given by

$$\boxed{h(s) = g(f(s))}$$

Thus the p.g.f. of the random sum is the **composite** (or **compound**) function of f and g.

Proof By the law of total probability

$$h_j = \Pr\{S_N = j\} = \sum_{n=0}^{\infty} \Pr\{N=n\} \Pr\{X_1 + X_2 + \cdots + X_n = j\}$$

$$= \sum_{n=0}^{\infty} g_n \Pr\{X_1 + X_2 + \cdots + X_n = j\}$$

But by the preceding corollary $X_1 + X_2 + \cdots + X_n$ has p.g.f. $f^n(s)$. Hence

$$h_j = \sum_{n=0}^{\infty} g_n \quad [\text{coefficient of } s^j \text{ in } f^n(s)].$$

Then,

$$h(s) = \sum_{j=0}^{\infty} h_j s^j$$

$$= \sum_{j=0}^{\infty} \sum_{n=0}^{\infty} g_n \quad [\text{coefficient of } s^j \text{ in } f^n(s)]s^j$$

$$= \sum_{n=0}^{\infty} g_n \sum_{j=0}^{\infty} \quad [\text{coefficient of } s^j \text{ in } f^n(s)]s^j$$

$$= \sum_{n=0}^{\infty} g_n f^n(s)$$

$$= g(f(s)).$$

This completes the proof.

Example

If the X_k are Bernoulli and N is Poisson with mean λ then the p.g.f. of X_1 is

$$f(s) = q + ps$$

whereas that of N is

$$g(s) = \sum_{k=0}^{\infty} e^{-\lambda} \frac{\lambda^k s^k}{k!} = e^{-\lambda(1-s)}.$$

Theorem 10.2 gives, for the p.g.f. of S_N,

$$h(s) = e^{\lambda(1-(q+ps))} = e^{-\lambda p(1-s)}.$$

Thus S_N is seen to be Poisson with mean λp (cf. Section 3.7).

10.5 THE PROBABILITY OF EXTINCTION

We now return to the Galton–Watson process, for which a pictorial representation of a realization is shown in Fig. 10.3. The process is the collection $\{X_0, X_1, X_2, \ldots\}$ for which we have the corresponding p.g.f.s defined by

$$P_n(s) = \sum_{k=0}^{\infty} \Pr\{X_n = k\} s^k, \qquad n = 0, 1, 2, \ldots \tag{10.6}$$

Since $X_0 = 1$ with probability one,

$$P_0(s) = s.$$

Also, we let the distribution of the number of descendants of an individual have p.g.f.

$$P(s) = \sum_{k=0}^{\infty} p_k s^k = P_1(s).$$

A recursion relation for the probability generating functions

We will prove the recursion relation

$$\boxed{P_n(s) = P(P_{n-1}(s))} \tag{10.7}$$

Proof Divide the nth generation of individuals into X_1 clans corresponding to the descendants of individuals in the first generation (see Fig. 10.3). Then we may write

$$X_n = X_{n,1} + X_{n,2} + \cdots + X_{n,X_1}$$

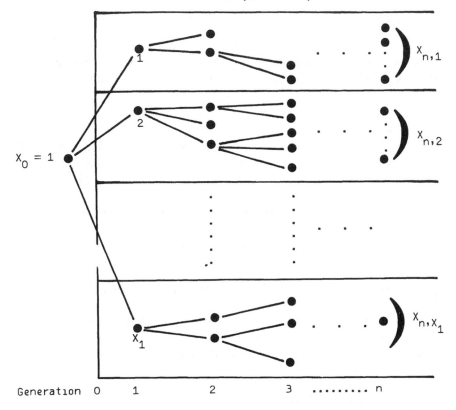

Figure 10.3 Representation of a sample path for a branching process showing X_1 descendants in the first generation.

where $X_{n,k}$ is the number of individuals in the nth generation with ancestor the kth individual in the first generation. But each $X_{n,k}$ is the number of individuals in a branching process starting with one individual and lasting for $n-1$ generations. Thus $X_{n,1}, X_{n,2}, \ldots$ are i.i.d. with the distribution of X_{n-1}. Applying Theorem 10.2, the result (10.7) quickly follows.

Extinction probability

If $p_0 = 1$ the population is extinct in the first generation and we will exclude this case. We also exclude the case $p_0 = 0$ because it leads to no possibility of extinction. The case $p_0 + p_1 = 1$ is also very easy to deal with (see Exercise 4). We therefore insist that

$$0 < p_0 \leqslant p_0 + p_1 < 1. \tag{10.8}$$

In general, extinction of the population occurs by the nth generation if

$X_n = 0$. We let the probability of this event be

$$x_n = \Pr\{X_n = 0\}, \qquad n = 1, 2, \dots \tag{10.9}$$

Notice that this does not imply that the population was not extinct before the nth generation. The probability that extinction occurs at any generation is

$$p_{\text{ext}} = \lim_{n \to \infty} x_n,$$

if this limit exists.

We will prove the following result for p_{ext}, essentially following Feller (1968, p. 295). Recall that μ is the mean number of individuals left behind in the next generation by any individual and $P(\cdot)$ is the corresponding p.g.f.

Theorem 10.3 In a Galton-Watson branching process, the probability of extinction is

$$p_{\text{ext}} = \begin{cases} 1, & \mu \leqslant 1, \\ x^*, & \mu > 1, \end{cases}$$

where x^* is the solution of $x = P(x)$ with magnitude less than one.

Proof We divide the proof into two parts. In part A we show that the extinction probability is a root of $x = P(x)$. The proof given is the usual one, although a much shorter one is given in the exercises. In part B we consider the various possibilities for μ.

A. *p_{ext} is a root of $x = P(x)$*

Putting $s = 0$ in (10.6) and using the definition (10.9) we have

$$x_n = P_n(0).$$

But according to (10.7), $P_n(s) = P(P_{n-1}(s))$ so, putting $s = 0$,

$$P_n(0) = P(P_{n-1}(0)).$$

Equivalently

$$x_n = P(x_{n-1}). \tag{10.10}$$

We now show that $\{x_n\}$ is an increasing sequence. We have

$$x_1 = p_0 = P(0)$$

and $p_0 > 0$ by assumption. Also, by (10.10),

$$x_2 = P(x_1).$$

But $P(s)$ is an increasing function for real non-negative s which can be seen by noting that $P'(s) = p_1 + 2p_2 s + \cdots$ is positive. Hence $P(x_1) > P(0)$; that is,

$$x_2 > x_1.$$

Similarly,

$$x_3 = P(x_2) > P(x_1) = x_2,$$

and so on. This establishes that $\{x_n\}$ is increasing. Each x_n is a probability, however, and must therefore be less than or equal to one. Thus $\{x_n\}$ is an increasing sequence whose terms are bounded above. Hence it is a convergent sequence (see theorems on sequences in any calculus text) and the limit

$$\lim_{n \to \infty} x_n = x$$

exists. From (10.10) the probability x, that extinction ever occurs, must satisfy the equation

$$\boxed{x = P(x)}, \qquad 0 \leqslant x \leqslant 1.$$

This completes part A of the proof.

B. *The cases* $\mu \leqslant 1$ *and* $\mu > 1$

We must have a root of $x = P(x)$ at $x = 1$ because $P(1) = 1$. We now show that if $\mu \leqslant 1$, $P(x)$ is always above x for $0 \leqslant x < 1$ whereas if $\mu > 1$ there is one root of $x = P(x)$ in $(0, 1)$.

By our assumption (10.8) there is a non-zero probability that an individual will have two or more descendants so we may write

$$P(s) = p_0 + p_{2+k} s^{2+k} + \sum_{\substack{m=1 \\ m \neq 2+k}}^{\infty} p_m s^m$$

where $k \geqslant 0$, $p_0 > 0$, $p_{2+k} > 0$ and $p_m \geqslant 0$ for all $m \geqslant 1$ excluding $m = 2 + k$. One can quickly check that this gives $P''(s) > 0$ for all $s > 0$. Using Taylor's theorem to expand P about $x = 1$,

$$P(1 - \varepsilon) = P(1) - \varepsilon P'(1) + \frac{\varepsilon^2}{2} P''(\xi), \qquad \varepsilon > 0,$$

where $1 - \varepsilon < \xi < 1$. Since $P'(1) = \mu$ from the properties of p.g.f.s (see exercises), and $P(1) = 1$, we get

$$P(1 - \varepsilon) = 1 - \mu\varepsilon + \frac{\varepsilon^2}{2} P''(\xi).$$

This leads to

$$P(1 - \varepsilon) - (1 - \varepsilon) = \varepsilon(1 - \mu) + \frac{\varepsilon^2}{2} P''(\xi). \tag{10.11}$$

If $\mu \leqslant 1$ the right side of (10.11) is always positive so $P(1 - \varepsilon) > 1 - \varepsilon$ and the curve $P(x)$ is always above x for $0 < x < 1$ as indicated in Fig. 10.4a. Thus the only root of $P(x) = x$ is $x = 1$ and the probability of extinction is 1.

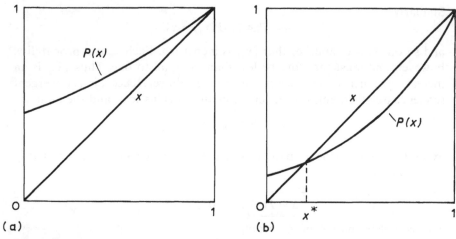

Figure 10.4 Solution of $P(x) = x$. In (a) where $\mu = 1$ the only root is $x = 1$ as $P(x) > x$ for $0 \leqslant x < 1$. In (b) $\mu > 1$ and $P(x) = x$ at $x^* < 1$ and at $x = 1$.

If $\mu > 1$, $P(1 - \varepsilon) - (1 - \varepsilon)$ is negative and hence $P(1 - \varepsilon) < 1 - \varepsilon$ for small enough ε. In fact there are two values of ε at which the right side of (10.11) is zero. The value $\varepsilon = 0$ corresponds to $x = 1$ and the other corresponds to $x = x^*$ as shown in Fig. 10.4b. (Note that by $x = 0$, $P(x)$ is above x because $P(0) = p_0 > 0$ by assumption.) It can also be seen graphically that the sequence $x_1, x_2, \ldots$ must converge to x^* (see Exercise 8). This completes the proof of Theorem 10.3.

This concludes our introductory treatment of branching processes. For more advanced treatments and related topics see Athreya and Ney (1970) and Jagers (1975, 1983).

REFERENCES

Athreya, K.B. and Ney, P. (1970). *Branching Processes*. Springer-Verlag, New York.

Brooks. R.F. (1981). Variability in the cell cycle and the control of proliferation. In *The Cell Cycle*. (ed. P.C.L. John). Cambridge University Press. Cambridge.

Carter, B.L.A. (1981). The control of cell division in *Saccharomyces cerevisiae*. In *The Cell Cycle*. (ed. P.C.L. John). Cambridge University Press, Cambridge.

Feller, W. (1968). *An Introduction to Probability Theory and its Applications*. Wiley, New York.

Galton, F. (1873). 'Problem 4001'. *Educational Times* (1 April 1873), 17.

Jagers, P. (1975). *Branching Processes with Biological Applications*. Wiley, London.

Jagers, P. (1983). Stochastic models for cell kinetics. *Bull. Math. Biol.*, **45** 507-19.

Kimball, J.W. (1984). *Cell Biology*. Addison-Wesley, Reading, Mass.

Prescott, D.M. (1976). *Reproduction of Eukaryotic Cells*. Academic Press, New York.

Steffensen, J.F. (1930). Om sandsynligheden for at afkommet ud dør. *Matematisk Tidsskrift*, B, **1**, 19-23.

Watson, H.W. and Galton, F. (1874). On the probability of extinction of families. *J. Anthropol. Inst. Gt. Brit. Ireland*, **4**, 138-44.

EXERCISES

1. Deduce from the recursion relation (10.1) that the variance of the population in the Galton-Watson process at generation n is

$$\text{Var}(X_n) = \begin{cases} \dfrac{\sigma^2 \mu^{n-1}(\mu^n - 1)}{\mu - 1}, & \mu \neq 1, \\ n\sigma^2, & \mu = 1. \end{cases}$$

2. Let X be a non-negative integer-valued random variable with probability generating function $f(s) = \sum_0^\infty f_k s^k$. Prove that

$$E(X) = f'(1)$$
$$\text{Var}(X) = f''(1) + f'(1) - f'^2(1).$$

3. Let $\{X_k, k = 1, 2, \dots\}$ be i.i.d. with $E(X_1) = \mu$ and $\text{Var}(X_1) = \sigma^2$ and let N be a Poisson random variable with parameter μ, independent of the X_k. Prove, using generating functions, that

$$S_N = X_1 + X_2 + \cdots + X_N,$$

has mean and variance given by

$$E(S_N) = \lambda\mu$$
$$\text{Var}(S_N) = \lambda(\mu^2 + \sigma^2).$$

4. Consider a branching process in which $X_0 = 1$ with probability one. Each individual leaves behind either zero descendants or one descendant with probabilities p_0 and p_1 respectively. Show that the probability of extinction at generation n is $p_0 p_1^{n-1}$. Sum the geometric series to show that the probability of extinction is $p_0/(1 - p_1)$. Obtain the same result by solving $P(x) = x$, where $P(x)$ is the generating function of the number of descendants.

5. A branching process is referred to as **binary fission** if an individual leaves either zero or two descendants. That is $p_2 = p$, $p_0 = 1 - p$, $0 < p < 1$. If $X_0 = 1$ with probability one, find the expectation and variance of the population size at generation n.

6. Viewing a branching process as a Markov chain, show that the transition probabilities for the binary fission case are

$$p_{jk} = \Pr\{X_{n+1} = k \mid X_n = j\} = \begin{cases} 0, & k \text{ odd} \\ \dbinom{j}{k/2} p^{k/2}(1 - p)^{j - k/2}, & k \text{ even}, \end{cases}$$

where $\dbinom{j}{i}$ is interpreted as zero if $i > j$.

7. For the binary fission branching process, solve the equation $P(x) = x$ to show that the probability of extinction is

$$P_{ext} = \frac{1 - \sqrt{1 - 4p(1 - p)}}{2p}.$$

8. Use Fig. 10.4b to show graphically that $x_n \to x^*$ when $\mu > 1$.

9. A branching process has initially one individual. Use the law of total probability in the form

$$\text{Pr(extinction)} = \sum_k \text{Pr(extinction}|k \text{ descendants) Pr}(k \text{ descendants})$$

to deduce that the extinction probability x is a solution of $x = P(x)$.

10. Let $\{X_n, n = 0, 1, 2, \ldots\}$ be a branching process with $X_0 = 1$ and with the number of offspring per individual $0, 1, 2$ with probabilities p, q, r, respectively, where $p + q + r = 1$ and $p, q, r > 0$. Show that if $q + 2r > 1$, the probability of extinction is

$$x^* = \frac{1 - q - \sqrt{(1 - q)^2 - 4pr}}{2r}.$$

11. Assume, very roughly speaking, that a human population is a branching process. What is the probability of extinction if the proportion of families having 0, 1 or 2 children are 0.2, 0.4 and 0.4 respectively?

11

Stochastic processes and an introduction to stochastic differential equations

11.1 DETERMINISTIC AND STOCHASTIC DIFFERENTIAL EQUATIONS

A differential equation usually expresses a relation between a function and its derivatives. For example, if $t \geqslant t_0$ represents time, and the rate of growth of a quantity $y(t)$ is proportional to the amount $y(t)$ already present, then we have

$$\frac{dy}{dt} = ky, \tag{11.1}$$

where k is a constant of proportionality. Equation (11.1) is called a **first-order differential equation** because the highest order derivative appearing is the first derivative. It is also called **linear** because both y and its derivative occur raised to power 1.

Equation (11.1) may be viewed as a prescription or **mathematical model** for finding y at all times subsequent to (or before) a given time t_0 at which the value y_0 of y is known. This is expressed in the solution of (11.1),

$$y(t) = y_0 e^{k(t - t_0)}, \tag{11.2}$$

which has the same form as the Malthusian population growth law of Section 9.1. It is also a formula for finding an asset value with compound interest when the initial value is y_0.

In the natural sciences (biology, chemistry, physics, etc.), differential equations have provided a concise method of summarizing physical principles. An important example of a **nonlinear first-order differential equation** is **Verhulst's logistic equation**:

$$\frac{dy}{dt} = ry\left(1 - \frac{y}{y^*}\right), \tag{11.3}$$

with $r \geqslant 0$. This equation is frequently used to model the growth of populations of organisms. The quantity y^* is called the **carrying capacity** whereas r is called the **intrinsic growth rate**. It will be seen in Exercise 1 that the solution of (11.3) which passes through the value y_0 at time $t = t_0$ is

$$y(t) = \frac{y^*}{\left[1 + \left(\dfrac{y^*}{y_0} - 1 \right) e^{-rt} \right]}. \tag{11.4}$$

Figure 11.1 shows how populations evolve for different starting values. As $t \to \infty$ the population approaches the value y^* asymptotically, which explains the term carrying capacity. Since its inception by the Belgian mathematician Verhulst (1838), the logistic equation has been used for many different populations, including those of cancer cells (Thompson and Brown, 1987) as well as human populations over countries (Pearl and Reed, 1920), continents and the world (Tuckwell and Koziol, 1992, 1993).

The differential equations (11.1) and (11.3) we have thus far considered are called **deterministic** because a given initial value **determines** the solution completely for **all** subsequent times. The behaviour of the solution is totally predictable and there are no chance elements. Put another way, the trajectory $y(t)$ is fixed (it is a particular function) and there are no haphazard or **random fluctuations**.

Deterministic differential equations proved to be extremely powerful in

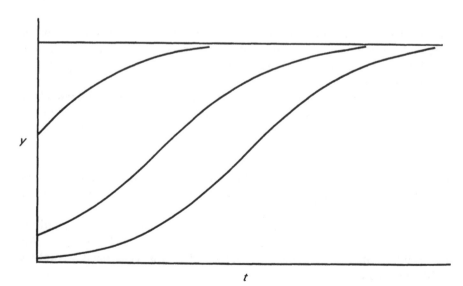

Figure 11.1 Showing solutions of a logistic differential equation for various initial population sizes.

some branches of classical physics and chemistry, but at the beginning of the twentieth century the study of atomic and subatomic systems indicated that deterministic theories were inadequate. Thus **quantum mechanics**, which is fundamentally probabilistic, was formulated to describe changes in very small systems (see for example, Schiff, 1955). Furthermore, in complex systems, containing millions or billions of interacting particles, the application of deterministic methods would have been so laborious that scientists also devised probabilistic methods for them. Such considerations for large collections of atoms or molecules lead to the discipline of **statistical mechanics** (see for example, Reichl, 1980).

In the latter part of the twentieth century quantitative methods have become increasingly widely used in the study of intrinsically complex systems such as arise in biology and economics. The use of deterministic methods is limited so there has been a large and rapid development in the application of probabilistic methods. One such very useful concept has been that of **stochastic differential equations**.

In the case of deterministic differential equations which are useful for quantitatively describing the evolution of natural systems, the solution is uniquely determined, usually by imposing a starting value and possibly other constraints. In the case of stochastic differential equations there are several possible trajectories or paths over which the system of interest may evolve. It is not known which of these trajectories will be followed, but one can often find the probabilities associated with the various paths. The situation

(A)   (B)

Figure 11.2a The three records on the left (A) show the fluctuations in the resting electrical potential difference across a nerve cell membrane. These fluctuations can be modelled with a stochastic differential equation involving a Wiener process - see section 12.7. On the right (B) is shown a histogram of amplitudes of the fluctuations, fitted with a normal density (from Jack, Redman and Wong, 1981).

Figure 11.2b Here are shown the fluctuations in the price of a share (Coles-Myer Limited) from week to week over a period of a few years. Such fluctuations can also be modelled using a stochastic differential equation – see section 12.7.

is similar to that in the simple random walk which we studied in Chapter 7, except that in most cases the time variable is continuous rather than discrete. We could say that the quantity we are looking at wanders all over the place in a random and thus unpredictable fashion.

Physical examples of quantities which might be modelled with stochastic differential equations are illustrated in Figs 11.2a and 11.2b. In the first of these we show a record of fluctuations in the electrical potential difference across the membrane of a nerve cell in a cat's spinal cord (a spinal motorneurone which receives messages from the brain and sends messages to a muscle fibre which may result in a movement). In the second example, the weekly variations in the price of an industrial share are shown from May 1990 to January 1993.

11.2 THE WIENER PROCESS (BROWNIAN MOTION)

The most useful stochastic differential equations have proven to be those which involve either Wiener processes or Poisson processes. When Wiener processes are involved, the solutions are usually continuous whereas when Poisson processes are involved the solutions exhibit jumps. Most of our discussion focuses on continuous processes so that our immediate concern is to define Wiener processes and discuss their properties.

In Section 7.8 we considered a simple random walk and let the step size get smaller as the rate of their occurrence increased. We took this to the limit of zero step sizes and an infinite rate of occurrence, but did so in such a way that the variance at any time neither vanished nor became unbounded. In fact, the variance of the limiting random process at time t was made to equal t.

The symbol we employ for the limiting process, which we call the **Wiener**

process, is $W = \{W(t), t \geq 0\}$. However, this process can be defined in a more general way, which makes no reference to limiting operations. In this section we will give this a more general definition, and discuss some of the elementary yet important properties of W.

Before we give this definition we will define a large class of processes to which both the Wiener process and Poisson process belong. This consists of those processes whose behaviour during any time interval is independent of their behaviour during any non-overlapping time interval. We will restrict our attention to processes whose index set (see section 7.1) is continuous.

Definition Let $X = \{X(t)\}$ **be a random process with a continuous parameter set** $[0, T]$, **where** $0 < T < \infty$. **Let** $n \geq 2$ **be an integer and suppose** $0 \leq t_0 < t_1 < t_2 < \cdots < t_n \leq T$. **Then** X **is said to be a random process with independent increments if the** n **random variables**

$$X(t_1) - X(t_0), X(t_2) - X(t_1), \ldots, X(t_n) - X(t_{n-1}),$$

are independent.

Thus, increments in X which occur in disjoint time intervals are independent. This implies that the evolution of the process after any time $s > 0$ is independent of the history up to and including s. Thus any process with independent increments is a **Markov process** as will be shown formally in the exercises. The converse is not true.

We have already encountered one example of an independent-increment process in section 9.2 – the **Poisson process**. Before defining a Wiener process, we mention that if the distributions of the increments of a process in various time intervals depend only on the lengths of those intervals and not their locations (i.e., their starting values), then the increments are said to be **stationary**. In section 9.2 we saw that for a Poisson process $N = \{N(t), t \geq 0\}$, the random increment $N(t_2) - N(t_1)$ is Poisson distributed with a parameter proportional to the length of the interval $(t_1, t_2]$. Thus a Poisson process has stationary independent increments.

Definition **A standard Wiener process** $W = \{W(t), t \geq 0\}$, **on** $[0, T]$, **is a process with stationary independent increments such that for any** $0 \leq t_1 < t_2 \leq T$, **the increment** $W(t_2) - W(t_1)$ **is a Gaussian random variable with mean zero and variance equal to** $t_2 - t_1$; **i.e.,**

$$E[W(t_2) - W(t_1)] = 0,$$
$$\mathrm{Var}[W(t_2) - W(t_1)] = t_2 - t_1.$$

Furthermore,

$$W(0) = 0,$$

with probability 1.

The probability density $p(x; t_1, t_2)$ of the increment of W in the interval (t_1, t_2) is defined through

$$\Pr\{W(t_2) - W(t_1) \in (x, x + \Delta x]\} = p(x; t_1, t_2)\Delta x + o(\Delta x).$$

From the definition of W we see that this is given by

$$p(x; t_1, t_2) = \frac{1}{\sqrt{(2\pi(t_2 - t_1))}} \exp\left[-\frac{x^2}{2(t_2 - t_1)}\right]. \tag{11.5}$$

In the case $t_1 = 0$, it is seen that the random variable $W(t_2)$ has mean 0 and variance t_2. Thus, **for any $t > 0$, $W(t)$ is a Gaussian random variable with mean 0 and variance t, so that its probability density $p(x; t)$ is given by the simple expression**

$$p(x; t) = \frac{1}{\sqrt{(2\pi t)}} \exp\left[\frac{-x^2}{2t}\right].$$

The word 'standard' in the definition refers to the fact that the mean is zero, the variance at t is t and the initial value is zero.

Sample paths

It can be proved for the random process defined above, that the **sample paths** or **trajectories** are continuous with probability one. Sample paths are also called **realizations** and correspond to a 'value' of the process when an experiment is performed. That is, supposing it is possible to observe a standard Wiener process over the time interval $[0, T]$, we would see, with probability one, a continuous function starting at the origin, wandering around haphazardly and reaching some random end-value $W(T)$ – as in Fig. 11.3.

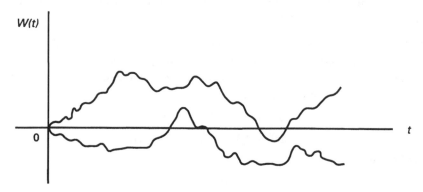

Figure 11.3 A depiction of a few sample paths for a Wiener process.

Note, however, that there are possibly discontinuous paths but these have zero probability associated with them. Usually, attention is restricted to those paths which are in fact continuous and in fact continuity of sample paths is often included in the definition. This is a convenient way to discard the problem of the discontinuous paths.

Although the probability of finding a continuous trajectory for W is one, *the probability is zero* that at any time $t \in [0, T]$ the path is differentiable. This is considered to be a pathological property and is one reason why a study of the Wiener process has been so interesting to mathematicians. This, and the fact that sample paths have *unbounded variation*, are proved and elaborated on in, for example, Hida (1980). An elementary consideration is given in Exercise 3.

Mean value and covariance function

An important property of a random process X is its mean at time t, $E(X(t))$, which is often called its **mean value function**, being a function of t alone. We have the mean and variance of $W(t)$ immediately from the above definition.

To further understand the behaviour of a random process, it is useful to know how its value at any time is connected with its value at any other time. Although knowing the joint probability distribution of these values would be nice, we may be content with a rougher indication. To this end we make the following definition.

Definition   The covariance function of a random process is the covariance (cf. Chapter 1) of the values of the process at two arbitrary times.

Note that sometimes the covariance function is called an *autocovariance function* to distinguish it from a covariance between two different processes.

It is also useful to define a class of processes whose covariance function depends only on the difference between the times at which it is evaluated and not on their location.

Definition   If the covariance function $Cov(X(s), X(t))$ depends only on $|t - s|$, the random process X is said to be covariance stationary. Other terms for this are wide-sense stationary or weakly stationary.

If X is a weakly stationary process, we may put

$$Cov(X(s), X(s + \tau)) = R(\tau).$$

We can see for such a process that (see Exercises):

(a) the mean value function is a *constant*; and,

(b) the covariance function is an *even function*:

$$R(\tau) = R(-\tau).$$

In the case of a standard Wiener process we will see that the following is true.

The covariance function of a standard Wiener process is

$$Cov(W(s), W(t)) = \min(s, t),$$

where min (.,.) is defined as the smaller of the two arguments.

Proof We utilize the fact that the increments of a Wiener process over disjoint (nonoverlapping) time intervals are independent random variables and hence have covariance equal to zero. With $s < t$ we have

$$Cov[W(s), W(t) - W(s)] = 0.$$

The quantity we seek can be written

$$Cov[W(s), W(t) + W(s) - W(s)].$$

But in general, if A, B, and C are three random variables (see Exercises),

$$Cov[A, B + C] = Cov[A, B] + Cov[A, C].$$

Thus,

$$Cov[W(s), W(t)] = Cov[W(s), W(t) - W(s)] + Cov[W(s), W(s)]$$
$$= Cov[W(s), W(s)]$$
$$= Var[W(s)] = s.$$

Had t been less than s we would have obtained t instead of s. Hence the covariance is the smaller of s and t, which proves the result.

Note that the Wiener process is not therefore covariance-stationary as the covariance of $W(s)$ and $W(t)$ depends directly on the magnitude of the smaller of s or t. For further information on the topics we have dealt with in this section, the reader may consult Papoulis (1965), Parzen (1962) and Yaglom (1973).

11.3 WHITE NOISE

Although the Wiener process is of central importance in the theory of stochastic differential equations, there is a useful related concept, called white noise, which we introduce in this section.

The paths traced out by a Wiener process are with probability one not differentiable. However, it is often convenient to talk about the derivative of

W as if it did exist. We use the symbol $w(t)$ for the 'derivative' of $W(t)$, and we call the random process $w = \{w(t), t \geqslant 0\}$, (Gaussian) **white noise**. However, it must be remembered that, strictly speaking, this process does not have a well-defined meaning – it is nevertheless heuristically useful.

The word *noise*, of course, refers to unwanted signals. If you are in a crowded cafeteria or football stadium or surrounded by dense city traffic, close your eyes and listen, you will hear a noise that seems an amorphous assortment of meaningless sounds; you generally won't be able to pick out particular signals unless they originate close-by. This kind of background noise is an acoustic approximation to white noise. Sound engineers have devices called white noise generators which are used to test the acoustic properties of rooms – the basic idea is to subject the chamber to all frequencies at once.

The mean value and covariance functions of white noise can be obtained from those of a Wiener process – as will be seen in the exercises. These turn out to be

$$E[w(t)] = 0,$$
$$\mathrm{Cov}[w(s), w(t)] = \delta(t - s). \qquad (11.6)$$

Thus the covariance is zero whenever $s \neq t$ and is very very large when $s = t$.

Covariance functions are often decomposed to see if there are regularities present, especially in the form of periodicities or harmonics of various frequencies. Such a decomposition is done using the following definition. Note that we restrict our attention to real-valued processes.

Definition   The spectral density $S(k)$ of a covariance-stationary random process whose covariance function is $R(t), t \geqslant 0$, is given by the integral

$$S(k) = \int_{-\infty}^{\infty} \cos(kt) R(t)\, \mathrm{d}t. \qquad (11.7)$$

The reader may recognize this as the *Fourier transform* of $R(t)$, recalling that the latter is here an *even* function of t. Another name for $S(k)$ is the **power spectrum** – it indicates the contributions from various frequencies to the total activity of the process.

A knowledge of the spectral density can be used to obtain the covariance function using the following **inversion formula** which is proved in courses of analysis (see for example Wylie, 1960),

$$R(t) = \frac{1}{2\pi} \int_{-\infty}^{\infty} S(k) \cos(kt)\, \mathrm{d}k. \qquad (11.8)$$

Let us see how various harmonics in $R(t)$ manifest themselves in $S(k)$. Suppose $S(k)$ were very much concentrated around the single frequency k_0

so we might put $S(k) = \delta(k - k_0)$. Then

$$R(t) = \frac{1}{2\pi} \int_{-\infty}^{\infty} \delta(k - k_0) \cos(kt)\, dk$$

$$= \frac{1}{2\pi} \cos(k_0 t),$$

where we have used the substitution property of the delta function (formula (3.13)). Thus we see that a very large peak in the spectral density $S(k)$ comes about at k_0 if there is a single dominant frequency $k_0/2\pi$ in the covariance function $R(t)$.

Let us consider white noise $w(t)$ from this point of view. We have from Equation (11.6), $R(\tau) = \delta(\tau)$. Substituting this in the definition of the spectral density gives

$$S(k) = \int_{-\infty}^{\infty} \cos(kt)\delta(t)\, dt$$

$$= 1,$$

where we have used the substitution property and the fact that $\cos(0) = 1$.

This tells us that the spectral density of white noise is a constant, independent of the frequency. That is, all frequencies contribute equally, from $-\infty$ to ∞, whereby we can see the analogy with 'white light'. Hence the description of the derivative of a Wiener process as (Gaussian) white noise. It is realized that it is not physically possible to have frequencies over such a huge range. In engineering practice white noise generators have cut-off frequencies at finite values - they are called **band-limited** white noises. Sometimes white noise is called **delta-correlated** noise.

11.4 THE SIMPLEST STOCHASTIC DIFFERENTIAL EQUATIONS - THE WIENER PROCESS WITH DRIFT

In this section we will take a first look at stochastic differential equations involving Wiener processes. A more detailed account will be given in the next chapter.

The increment in a standard Wiener process in a small time interval $(t, t + \Delta t]$ is

$$\Delta W(t) = W(t + \Delta t) - W(t),$$

and we know from above that ΔW is normally distributed with mean zero and variance Δt. We use a similar notation as in differential calculus and use the symbol $dW(t)$ or dW to indicate the limiting increment or **stochastic differential** as $\Delta t \to 0$. The simplest stochastic differential equation involving a Wiener process is thus:

$$dX = dW \qquad (11.9)$$

which states that the increments in X are those of W. The solution of (11.9) is

$$X(t) = X(0) + W(t),$$

which states that the value of the process X at time t, namely the random variable $X(t)$, is equal to the sum of two random variables: the initial value $X(0)$ and the value of a standard Wiener process at time t. Equation (11.9) is interpreted more rigorously as the corresponding integral

$$\int_0^t dX(t') = \int_0^t dW(t'),$$

whose meaning will be explained in section 12.5. This gives

$$X(t) - X(0) = W(t) - W(0) = W(t),$$

which is the same as (11.10) because from the definition, $W(0) = 0$, identically.

Notice that when writing stochastic differential equations involving a Wiener process, we usually avoid writing time derivatives because, as we have seen, these do not, strictly speaking, exist. However, we can, if we are careful in our interpretation, just as well write (11.9) as

$$\frac{dX}{dt} = w(t),$$

where w is white noise.

We may perform simple algebraic operations on a standard Wiener process. For example, we can form a new process whose value at time t is obtained by multiplying $W(t)$ by a constant σ, usually assumed to be positive; adding a linear function of time μt, where μ can be negative, zero, or positive; and giving a particular initial value $X(0) = x_0$:

$$X(t) = x_0 + \mu t + \sigma W(t) \qquad (11.10)$$

This defines a *Wiener process with drift* μt and *variance parameter* σ. The drift function μt here is linear, though any other deterministic function of time can be added instead of μt. For the random process defined by (11.10) we write the *stochastic differential equation*

$$dX = \mu\, dt + \sigma\, dW, \qquad (11.11)$$

and say that (11.10) is a solution of (11.11) with a particular initial value.

The following properties of a Wiener process with drift will be verified in the exercises:

$$E[X(t)] = x_0 + \mu t,$$
$$\mathrm{Cov}[X(s), X(t)] = \sigma^2 \min(s, t).$$
$$\mathrm{Var}[X(t)] = \sigma^2 t.$$

To obtain the probability density function for the Wiener process with drift, as defined by (11.10), we note, as proven in introductory probability theory, that linear operations on a Gaussian random variable produce another Gaussian random variable. Thus $X(t)$ must be a Gaussian random variable with mean and variance as given above. Its probability density function, conditioned on an initial value x_0, is defined through either

$$p(x, t \mid x_0) = \lim_{\Delta x \to 0} \frac{\Pr\{x < X(t) < x + \Delta x \mid X(0) = x_0\}}{\Delta x},$$

or

$$p(x, t \mid x_0)\Delta x = \Pr\{x < X(t) < x + \Delta x\} + o(\Delta x),$$

where $t > 0$, $-\infty < x_0, x < \infty$. This density must be given by

$$p(x, t \mid x_0) = \frac{1}{\sqrt{2\pi\sigma^2 t}} \exp\left[-\frac{(x - x_0 - \mu t)^2}{2\sigma^2 t} \right]. \tag{11.12}$$

(Note that when dealing with continuous random variables as we are here, we can put $<$ rather than $\leqslant$ in inequalities because single points make no contribution.)

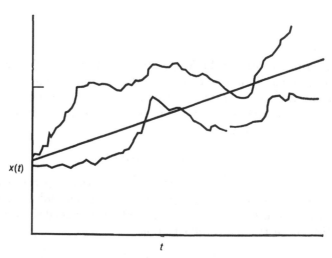

Figure 11.4 A depiction of a few sample paths for a Wiener process with drift $X(t) = x_0 + \mu t + \sigma W(t)$ with $x_0 = 1$, $\mu = \frac{1}{2}$, and $\sigma = 1$.

In anticipation of the material in the next chapter, we mention that the function $p(x, t | x_0)$, given in (11.12), satisfies a simple partial differential equation called a heat equation. This will be familiar to students either from calculus or physics courses and here takes the form,

$$\frac{\partial p}{\partial t} = -\mu \frac{\partial p}{\partial x} + \frac{\sigma^2}{2} \frac{\partial^2 p}{\partial x^2} \tag{11.13}$$

as will be verified in the exercises. It can be seen, therefore, that asserting that the probability density of a Markov process satisfies this partial differential equation, is, for all intents and purposes, the same as saying that the process is a Wiener process with drift.

Figure 11.4 illustrates how a Wiener process with drift might behave in the case of a positive drift, with drift parameter $\mu = \frac{1}{2}$ and variance parameter $\sigma = 1$ when $x_0 = 1$.

11.5 TRANSITION PROBABILITIES AND THE CHAPMAN-KOLMOGOROV EQUATION

Before considering a wide class of random processes which can be succinctly described in the language of stochastic differential equations, we will lay the groundwork for an analytical approach to studying their properties.

We saw in Chapter 8 that the fundamental descriptive quantity for Markov chains in discrete time was the set (or matrix) of transition probabilities. For the processes we considered, it was sufficient to specify the one-step transition probabilities, as the probabilities of all other transitions could be obtained from them. In particular, if the initial probability distribution was specified, the probability distribution of the process could be obtained at any time point - see Equation (8.11). Similarly, in Chapter 9, we saw that a set of transition probabilities could be used to quantitatively describe the evolution of Markov chains in continuous time.

The processes we are concerned with here are **Markov processes in continuous time which take on a continuous set of values.** The evolution of such processes is also specified by giving a set of transition probabilities as alluded to in the case of a Wiener process with drift. In general, let $\{X(t), t \geqslant 0\}$ be such a process. Then the **transition probability distribution function** gives the probability distribution of the value of the process at a particular time, conditioned on a known value of the process at some earlier time.

Definition   Let X be a continuous time random process taking on a continuous set of values. The transition probability distribution function $P(y, t | x, s)$, with $s \leqslant t$, is the distribution function of $X(t)$ conditioned on the event $X(s) = x$.

Thus,

$$P(y, t | x, s) = \Pr\{X(t) \leqslant y | X(s) = x\},$$

where usually $0 \leqslant s \leqslant t \leqslant \infty$ **and** $-\infty < x, y < \infty.$

The variables (x, s) which refer to the state of affairs at the earlier time are called *backward variables*, whereas those which refer to the later time are called *forward variables*.

Furthermore, if $P(y, t | x, s)$ is differentiable with respect to y, then its derivative, with t, x and s fixed,

$$p(y, t | x, s) = \frac{\partial P}{\partial y},$$

is called the **transition probability density function** (tpdf) of the process X.

The Chapman–Kolmogorov integral equation

In section 8.5 the *Chapman–Kolmogorov equations* were established for discrete-time Markov chains. These equations imply that a transition from one state to another in more than one time step must involve an intermediate passage to one of the permissible values of the process. For the type of

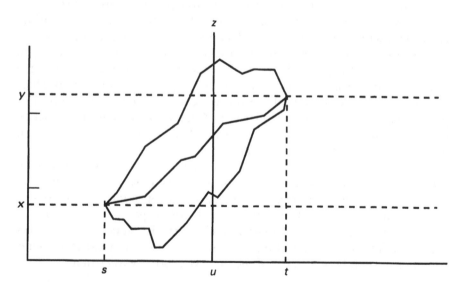

Figure 11.5 Showing how passage from (x, s) to (y, t) must involve passage to one of a continuum of intermediate values, z, at some intermediate time, u. Summing (integrating) over all possible paths gives the Chapman–Kolmogorov equation (11.14).

process of concern to us here, the corresponding equation involves not a sum but an integral over intermediate possible values, reflecting the fact that the state-space is continuous. Refer to Fig. 11.5.

We see that a transition from state x at time s to state y at a later time t, with associated probability density $p(y, t | x, s)$, must, at some intermediate time $u \in (s, t)$, involve passage to one of the permissible values z, here assumed to be any real number. The probability of a transition from (x, s) to (z, u) followed by a transition from (z, u) to (y, t) is proportional to the product $p(z, u | x, s) p(y, t | z, u)$. Integrating over these intermediate values gives the **Chapman–Kolmogorov equation**,

$$p(y, t | x, s) = \int_{z = -\infty}^{z = \infty} p(y, t | z, u) p(z, u | x, s) \, dz. \tag{11.14}$$

It can be seen that this is an integral over all possible paths from (x, s) to (y, t). The Chapman–Kolmogorov equation is useful for deriving differential equations satisfied by the transition density function.

Using similar reasoning, we may find the (absolute) probability distribution of the process at time t from a knowledge of the initial distribution and the transition probability density function or the transition probability distribution function. Let $f(x)$, where $-\infty < x < \infty$, be the density of $X(0)$. Then, to get the probability of being in state y at $t > 0$, we have to integrate over all possible initial values x, weighted with $f(x) \, dx$ and with the probability of a transition from x to y:

$$\Pr\{y < X(t) < y + dy\} = \left\{ \int_{x = -\infty}^{x = \infty} f(x) p(y, t | x, 0) \, dx \right\} dy.$$

(Note that this is *not* a conditional probability.)

Similarly, the distribution function of $X(t)$ is given by

$$\Pr\{X(t) \leqslant y\} = \int_{z = -\infty}^{z = y} \left(\int_{-\infty}^{\infty} f(x) p(z, t | x, 0) \, dx \right) dz$$

$$= \int_{-\infty}^{\infty} f(x) \left(\int_{-\infty}^{y} p(z, t | x, 0) \, dz \right) dx.$$

That is,

$$\Pr\{X(t) \leqslant y\} = \int_{-\infty}^{\infty} f(x) P(y, t | x, 0) \, dx,$$

where P is the transition probability distribution function.

In the special case, often encountered, where the initial value is not random but a particular specified value x_0, say, so that $f(x) = \delta(x - x_0)$, we have, by the substitution property of the delta function,

$$\Pr\{X(t) \leqslant y\} = \int_{-\infty}^{\infty} \delta(x - x_0) P(y, t \,|\, x, 0) \, \mathrm{d}x$$

$$= P(y, t \,|\, x_0, 0),$$

as expected. Thus the absolute and transition probability distributions coincide.

When one is seeking the properties of a process X, one may work with the random variables $\{X(t), t \geqslant 0\}$ directly, or one may work with the transition probability functions. The latter approach is called the *analytical approach* and is often more useful than the direct approach because it involves solving differential equations, which is a long and much-studied discipline. The direct approach usually involves stochastic integrals which we shall consider in section 12.5.

REFERENCES

Hida, T. (1980). *Brownian Motion.* Springer-Verlag, New York.

Jack, J.J.B., Redman, S.J. and Wong, K. (1981). The components of synaptic potentials evoked in cat spinal motoneurones by impulses in single group Ia afferents. *J. Physiol.* **321**, 65–96.

Papoulis, A. (1965). *Probability. Random Variables and Stochastic Processes.* McGraw-Hill, New York.

Parzen, E. (1962). *Stochastic Processes.* Holden-Day, San Francisco.

Pearl, R. and Reed, L.J. (1920). On the rate of growth of the population of the United States since 1790 and its mathematical representation. *Proc. Natl. Acad. Sci. USA*, **6**, 275–288.

Reichl, L.E. (1980). *A Modern Course in Statistical Physics.* University of Texas, Austin.

Schiff, L.I. (1955). *Quantum Mechanics.* McGraw-Hill, New York.

Thompson, J.R. and Brown, B.W. (eds) (1987). *Cancer Modeling.* Marcel Dekker, New York.

Tuckwell, H.C. and Koziol, J.A. (1992). World population. *Nature,* **359**, 200.

Tuckwell, H.C. and Koziol, J.A. (1993). World and regional populations. *BioSystems* **31**, 59–63.

Verhulst, P.F. (1838). Notice sur la loi que la population suit dans son accroissement. *Corr. Math. Phys.*, **10**, 113–121.

Wylie, C.R. (1960). *Advanced Engineering Mathematics.* McGraw-Hill, New York.

Yaglom, A.M. (1973). *An Introduction to the Theory of Stationary Random Functions.* Dover, New York.

EXERCISES

1. Prove that the solution of the logistic differential equation (11.3) is in fact given by (11.4). (*Hint*: Separate the variables by putting the equation in the form $f(y)\,\mathrm{d}y = g(t)\,\mathrm{d}t$, and integrate each side.)

2. Show that a continuous time process with independent increments is a Markov process. (*Hint:* It will suffice to proceed as in Exercise 7.2; examine $\Pr(X(t_3) = z \mid X(t_2) = y, X(t_1) = x)$, where $t_1 < t_2 < t_3$.)

3. Let $\Delta W = W(t + \Delta t) - W(t)$ be an increment in a standard Wiener process. Examine the limits as $\Delta t \to 0$ of (i) $E[\Delta W]$, (ii) $\mathrm{Var}[\Delta W]$, (iii) $E[\Delta W/\Delta t]$, and (iv) $\mathrm{Var}[\Delta W/\Delta t]$ to provide an indication that W has continuous but not differentiable sample paths.

4. Show that if A, B, and C are three random variables,

$$\mathrm{Cov}[A,(B+C)] = \mathrm{Cov}[A, B] + \mathrm{Cov}[A, C].$$

5. Prove that if a and b are constants, and X and Y are random variables, then

(a) $\mathrm{Cov}[X + a, Y + b] = \mathrm{Cov}[X, Y]$;

(b) thus $\mathrm{Cov}[X, Y] = \mathrm{Cov}[X - E[X], Y - E[Y]]$, so means can always be subtracted when calculating covariances;

(c) $\mathrm{Cov}[a, X, bY] = ab\,\mathrm{Cov}[X, Y]$.

6. Show that for a covariance (weakly) stationary process, the mean value function is constant and the covariance function satisfies $R(\tau) = R(-\tau)$.

7. The derivative $\{X'(t)\}$ of a random process $\{X(t)\}$ can be defined in the usual way by

$$X'(t) = \lim_{\Delta t \to 0} \frac{X(t + \Delta t) - X(t)}{\Delta t}.$$

Use the results

$$E[X'(t)] = \frac{d}{dt} E[X(t)],$$

$$\mathrm{Cov}[X'(s), X'(t)] = \frac{\partial^2}{\partial s \partial t}(\mathrm{Cov}[X(s), X(t)]),$$

to verify the given formulas for the mean and covariance function of white noise.

8. Establish the following results for a Wiener process with drift μt and variance parameter σ:

$$E[X(t)] = x_0 + \mu t,$$
$$\mathrm{Var}[X(t)] = \sigma^2 t,$$
$$\mathrm{Cov}[X(s), X(t)] = \sigma^2 \min(s, t).$$

9. Compute the probability that the Wiener process with drift $X = 3W + t$ satisfies the inequality

$$0 \leqslant X(t) \leqslant 1.$$

10. Find functions $f_1(t)$ and $f_2(t) = -f_1(t)$ such that a standard Wiener process is between f_1 and f_2 with probability (a) 0.5, (b) 0.95.
11. What is the probability that $-\sqrt{t} < W(t) < \sqrt{t}$?
12. Prove that the transition probability density (11.12) of a Wiener process with drift satisfies the heat equation (11.13).
13. Use Theorem 6.5 to find the characteristic function of $X(s) = x_0 + \mu s + \sigma W(s)$.
14. Let $N = \{N(t)\}$ be a Poisson process with parameter λ. Find the mean and covariance functions of N.
15. Let $M = dN/dt$, where N is as in Exercise 14. What would a sample path of M look like? Use the results of Exercise 7 to ascertain the mean and covariance functions of M.
16. For $s > 0, t > 0$ find the *correlation coefficient* $\rho(s, t)$ (see section 1.3) of $W(s)$ and $W(t)$. Assume $s < t$ and s is fixed. What happens to ρ as $t \to \infty$?

12

Diffusion processes, stochastic differential equations and applications

12.1 DIFFUSION PROCESSES AND THE KOLMOGOROV (OR FOKKER-PLANCK) EQUATIONS

To introduce a wide class of random processes with properties similar to those of a Wiener process with drift, we generalize the constant *drift parameter* μ and *variance parameter* σ of such a process, so that they vary with the value of the process and possibly the time. For a general process X, we have that the *increment* in the small time interval $(t, t + \Delta t]$ is

$$\Delta X(t) = X(t + \Delta t) - X(t).$$

Now the properties of this increment may depend on the time t and the value x of the process at the beginning of the small time interval. We therefore condition on $X(t) = x$ and define the **infinitesimal first moment**, or **infinitesimal mean**, as

$$\alpha(x, t) = \lim_{\Delta t \to 0} \frac{E[\Delta X(t)| X(t) = x]}{\Delta t}. \tag{12.1}$$

Note that because we have taken the expectation, this is not a random quantity. Thus $\alpha(x, t)$ is a deterministic function of x and t. Similarly we define the **infinitesimal second moment**, or, as will be seen in Exercise 1, **infinitesimal variance**,

$$\beta(x, t) = \lim_{\Delta t \to 0} \frac{E[(\Delta X)^2| X(t) = x]}{\Delta t}. \tag{12.2}$$

We assume that the higher order infinitesimal moments are zero, so that, for $n = 3, 4, \ldots$,

$$\lim_{\Delta t \to 0} \frac{E[(\Delta X)^n| X(t) = x]}{\Delta t} = 0. \tag{12.3}$$

This indicates that changes in the process in small time intervals will be small, and in fact small enough to make the sample paths continuous for suitably chosen functions α and β. Such a process is called a **diffusion process** and behaves in a fashion similar to a Wiener process – although its paths are continuous, they are with probability one non-differentiable. The **drift** (α) and **diffusion** (β) components depend on the position and the time.

Once the drift and diffusion terms are specified, we are in a position to obtain as much information as we require about the process if we can find its transition probability density function. Fortunately this can always be done because as the following theorem indicates, this function satisfies a partial differential equation which is a general form of the much studied heat equation – the differential equation (11.13) satisfied by the transition probability functions of the Wiener process with drift.

Theorem 12.1 Let $p(y, t | x, s)$ be the transition probability density function for a diffusion process with first and second infinitesimal moments $\alpha(y, t)$ and $\beta(y, t)$ as defined in equations (12.1) and (12.2) respectively. Then p satisfies the forward Kolmogorov equation

$$\frac{\partial p}{\partial t} = -\frac{\partial(\alpha p)}{\partial y} + \frac{1}{2}\frac{\partial^2(\beta p)}{\partial y^2},$$ (12.4)

with suitable initial and boundary conditions.

Equation (12.4) is also called a *Fokker-Planck equation*, especially by physical scientists, who sometimes refer to it as a 'Master equation', to emphasize its generality. Proof that this equation follows from the Chapman–Kolmogorov equation (11.14) and the relations (12.1)-(12.3), though not difficult, is rather long and is hence omitted here. Interested readers may refer to, for example, Jaswinski (1970).

The equation (12.4) is called the *forward equation* because the variables x and s which refer to the earlier event are considered to be fixed as the later variables y and t vary. One may also consider p as a function with fixed values of y and t, and allow the earlier variables x and s to vary. This gives rise to the *backward equation* which is often very useful, for problems such as ascertaining times at which a certain value or set of values is first attained.

Theorem 12.2 Let α and β be the first and second infinitesimal moments of a diffusion process. If the process has a transition probability density function $p(y, t | x, s)$, then this density considered as a function of x and s with y and t fixed, satisfies the backward Kolmogorov equation,

$$-\frac{\partial p}{\partial s} = \alpha\frac{\partial p}{\partial x} + \frac{1}{2}\beta\frac{\partial^2 p}{\partial x^2}.$$ (12.5)

The derivation of the backward equation from the Chapman-Kolmogorov equation is also relatively straightforward but will again not be given here. In addition it will be seen that the transition probability distribution function $P(y, t|x, s)$ also satisfies equation (12.5).

Time-homogeneous processes

In many problems of physical interest, the behaviour of a process depends not on the actual value of the time, but rather the *length* of the time interval since the process was *switched on*. Such a process is called **temporally (or time-) homogeneous** and nearly all diffusion processes which have arisen in applications fall into this category. (Note that some authors refer to such a process as one with stationary transition probabilities.) Clearly the first and second infinitesimal moments of such a process do not depend explicitly on time, so we have $\alpha(x, t) = \alpha(x)$, and $\beta(x, t) = \beta(x)$, being functions only of the state variable. Furthermore we have

$$p(y, t|x, s) = p(y, t - s|x, 0),$$

so that we can conveniently drop one of the arguments of the transition density. Thus we can use $p(y, t|x)$ for the density associated with transitions from a state $X(0) = x$. That is,

$$p(y, t|x) = \frac{\partial}{\partial y} \Pr\{X(t) \leqslant y | X(0) = x\}.$$

The forward and backward Kolmogorov equations now take somewhat simpler forms. For the *forward equation* we have

$$\frac{\partial p}{\partial t} = -\frac{\partial(\alpha(y)p)}{\partial y} + \frac{1}{2}\frac{\partial^2(\beta(y)p)}{\partial y^2}, \tag{12.6}$$

and, as is seen in the exercises, the *backward equation* simplifies to

$$\frac{\partial p}{\partial t} = \alpha(x)\frac{\partial p}{\partial x} + \frac{1}{2}\beta(x)\frac{\partial^2 p}{\partial x^2}. \tag{12.7}$$

Boundary conditions

When investigating the properties of a diffusion process by means of the Kolmogorov differential equations, it is necessary to prescribe appropriate boundary conditions in order to solve the latter. To be specific, let us assume that the diffusion process is on the interval (x_1, x_2) and the time at which it commences is $t = 0$. Assume from now on also that the process is **time-homogeneous**. For the Kolmogorov equations involving the transition density

$p(y, t | x)$, the initial condition must take the form:

$$p(y, 0 | x) = \delta(y - x),$$

because $X(0) = x$ with probability one.

The boundaries of intervals of definition of diffusion processes were investigated by Feller (1952). There are four main classes of boundary, exemplified in what follows by considering x_2 to be a boundary point. The definitions will be given here, together with results on establishing the nature of a boundary point. Examples will be given in sections 12.3 and 12.4.

(1) Regular boundary: At a regular boundary point one may choose certain boundary conditions to be satisfied by the transition density. Usually the condition applied will be either

(a) absorbing barrier – at which the process X is made to terminate, the mathematical requirement being

$$p(x_2, t | x) = 0;$$

or (b) reflecting barrier – at which the process is reflected back into the interior of the interval (x_1, x_2). Now, the quantity

$$J(y, t) = \alpha p - \frac{1}{2} \frac{\partial}{\partial y} (\beta p),$$

can be shown to be a **probability current**; i.e., the rate of flow of probability mass across the value or position y – see for example Stratonovich (1963). Hence if x_2 is a reflecting barrier, we must have the boundary condition

$$\left[\alpha p - \frac{1}{2} \frac{\partial}{\partial y} (\beta f) \right]_{y = x_2} = 0.$$

(2) Natural boundary: Such a boundary is unattainable by the process in a finite time and the transition probability density must be zero there.

(3) Exit boundary: An exit boundary is intrinsically absorbing so the transition density must vanish at such a point. However, there is no choice as in the case of a regular boundary with an imposed absorbing condition. The boundary is absorbing by virtue of the nature of the first two infinitesimal moments.

(4) Entrance boundary: At such a boundary point, probability mass actually flows in to the interval of definition of the process. The situation is similar to that at an exit boundary, in the sense that boundary conditions cannot be arbitrarily chosen but are forced on solutions by virtue of the nature of the drift and diffusion components.

In determining the nature of the boundaries for a given diffusion process, the main idea is that the nature of the boundaries and hence the kinds of boundary conditions that can or must be employed, depends on the behaviour of the first and second infinitesimal moments $\alpha(y)$ and $\beta(y)$ at x_1 and x_2. The results may be summarized as follows.

Starting with the functions α and β, which are always given, and assumed from now on to be continuous, we define the following further four functions.

$$f(x) = \exp\left[-\int^x \frac{\alpha(y)}{\beta(y)} \, dy \right],$$

$$g(x) = \frac{1}{\beta(x)f(x)},$$

$$h(x) = f(x) \int^x g(y) \, dy,$$

and

$$k(x) = g(x) \int^x f(y) \, dy.$$

To examine a boundary point, x_2 for example, we examine whether the integrals of the four functions f, g, h and k are finite on (x', x_2), where $x' \in (x_1, x_2)$. Let us introduce the following notation:

$$I_f(x_2) = \int_{x'}^{x_2} f(z) \, dz$$

$$I_g(x_2) = \int_{x'}^{x_2} g(z) \, dz$$

etc.

The nature of the boundaries can be established as follows:

(1) If $I_f(x_2) < \infty$ and $I_g(x_2) < \infty$, then x_2 is a **regular boundary**.
(2) If $I_g(x_2) = \infty$ and $I_k(x_2) < \infty$, then x_2 is an **exit boundary**.
(3) If $I_g(x_2) < \infty$ and $I_k(x_2) < \infty$, then x_2 is an **entrance boundary**.
(4) Otherwise, x_2 is a **natural boundary**.

Similarly one determines the nature of the other boundary point x_1 by considering integrals of the type

$$\int_{x_1}^{x'} f(z) \, dz.$$

The above results on boundary classification may be obtained by considering quantities such as exit or escape times for the given process, the theory of which depends on the backward Kolmogorov equation. Of course when solving either (12.4) or (12.5), solutions must be nonnegative and bounded. Although the above calculations may seem complicated, some experience will enable insights into the nature of boundary points to be established with a small amount of work or possibly by inspection of the functions α and β.

12.2 STATIONARY DISTRIBUTIONS

We saw in Chapter 8 that, for some Markov chains, a steady state distribution was approached as the number of time-steps increased. Physically this means that after (perhaps) a very long time, the probability of finding the system in a given state is a constant, regardless of the initial value or initial probability distribution of the process. The steady state distribution is time-independent and is for that reason often referred to as an *invariant distribution*. Many Markov chains do not possess stationary distributions, or may not approach them, as we saw in sections 8.7 and 8.12.

A similar situation occurs for time-homogeneous diffusion processes. For these, if a stationary distribution is attained as $t \to \infty$, the corresponding transition probability density function $p(y, t \mid x)$ must become independent of t and lose its dependence on x, the initial value $X(0)$ of the process. That is, it is a function of y alone.

Let us denote the corresponding **stationary density** by $p^*(y)$. This density must satisfy the forward Kolmogorov equation (12.6) with the time derivative set to zero. This leads to the following result, in which we assume that a diffusion process $\{X(t), t \geq 0\}$ takes values in the open interval (x_1, x_2).

Theorem 12.3   Let X be a time-homogeneous diffusion process with first and second infinitesimal moments $\alpha(y)$ and $\beta(y)$, respectively. If the probability distribution of $X(t)$ becomes stationary as $t \to \infty$, then the corresponding density $p^*(y)$ is given by

$$p^*(y) = \frac{1}{\beta(y)\phi(y)} \left[2K_1 \int^y \phi(z)\,dz + K_2 \right], \tag{12.8}$$

where

$$\phi(y) = \exp\left[2 \int^y \frac{\alpha(z)}{\beta(z)}\,dz \right], \tag{12.9}$$

and where K_1 and K_2 are constants determined by the boundary conditions and the normalization requirement

$$\int_{x_1}^{x_2} p^*(y)\,dy = 1. \tag{12.10}$$

Proof Setting the time-derivative in (12.6) equal to zero gives the following ordinary differential equation for $p^*(y)$:

$$\frac{1}{2}\frac{d^2}{dy^2}(\beta p^*) - \frac{d}{dy}(\alpha p^*) = 0. \tag{12.11}$$

One may immediately perform a y-integration, the right hand side yielding a constant K_1, say. Thus p^*: satisfies,

$$\frac{1}{2}\frac{d}{dy}(\beta p^*) - \alpha p^* = K_1. \tag{12.12}$$

Putting

$$\beta p^* = q,$$

this may be written

$$\frac{dq}{dy} - \frac{2\alpha}{\beta}q = 2K_1, \tag{12.13}$$

which is in the standard form of a first-order linear differential equation as encountered in calculus courses – see for example Berkey (1984). An *integrating factor* is given by

$$\phi = \exp\left[\int^y \frac{2\alpha(y')}{\beta(y')}dy'\right],$$

with derivative

$$\frac{d\phi}{dy} = \frac{2\alpha}{\beta}\exp[.]$$

$$= \frac{2\alpha\phi}{\beta}.$$

Multiplying equation (11.27) by ϕ thus gives (product rule),

$$\frac{d}{dy}[\phi q] = 2K_1\phi.$$

An integration yields

$$\phi q = \int^y 2K_1\phi(y')dy' + K_2,$$

where K_2 is a second constant of integration. Rearranging this gives the desired result, (12.8).

Now if the random process X is restricted to the interval (x_1, x_2), which is the case if, for example, the boundaries are reflecting or natural, then the probability current must be zero at the boundary points. As can be seen from Equation (12.12), K_1 is then necessarily zero and the formula for the stationary density simplifies to

$$p^*(y) = \frac{K_2}{\beta(y)\phi(y)}.$$

Furthermore, since $y\in(x_1,x_2)$ and $\int p^*\,dy = 1$, we have explicitly

$$K_2 = \frac{1}{\displaystyle\int_{x_1}^{x_2} \frac{dy}{\beta(y)\phi(y)}}.$$

In certain other cases, the constants K_1 and K_2 can be found by using the normalization condition (11.24) and the necessary condition

$$p^*(y) \geqslant 0,\, y\in(x_1,x_2).$$

12.3 THE WIENER PROCESS WITH DRIFT

Infinitesimal moments and Kolmogorov equations

For the process defined in Equation (11.10), where

$$X(t) = x_0 + \mu t + \sigma W(t),$$

we have

$$\Delta X = \mu\Delta t + \sigma\Delta W.$$

Since Δt and ΔW are both independent of the value of $X(t)$, the first infinitesimal moment α is, from the definition (12.1),

$$\alpha(x,t) = \lim_{\Delta t \to 0} E\left[\frac{\Delta X}{\Delta t}\right],$$

$$= \lim_{\Delta t \to 0} E\left[\frac{\mu\Delta t}{\Delta t}\right] = \mu,$$

because $E[\Delta W] = 0$.

For the second infinitesimal moment we have (see Exercise 1),

$$\beta(x,t) = \lim_{\Delta t \to 0} \frac{\text{Var}[\Delta X \mid X(t) = x]}{\Delta t}$$

$$= \lim_{\Delta t \to 0} \frac{\sigma^2\Delta t}{\Delta t} = \sigma^2.$$

The process is time-homogeneous because α and β do not depend on time.

The forward Kolmogorov equation (12.6) becomes

$$\frac{\partial p}{\partial t} = -\mu \frac{\partial p}{\partial y} + \frac{\sigma^2}{2} \frac{\partial^2 p}{\partial y^2}, \tag{12.14}$$

and the backward equation (12.7) becomes

$$\frac{\partial p}{\partial s} = \mu \frac{\partial p}{\partial x} + \frac{\sigma^2}{2} \frac{\partial^2 p}{\partial x^2}.$$

Boundary classification

Since $\alpha(y) = \mu$ and $\beta(y) = \sigma^2$, we have, for the function f of section 12.1, on using an arbitrary lower limit for convenience,

$$f(x) = \exp\left[-\int_v^x \frac{2\mu}{\sigma^2} \, dy \right]$$

$$= C \exp\left[-2\frac{\mu}{\sigma^2} x \right],$$

where C is a constant. We see therefore that provided x_1 and x_2 are finite then so too are the quantities $I_f(x_1), I_f(x_2), I_g(x_1)$ and $I_g(x_2)$. Hence *any finite boundary point is* **regular** *for a Wiener process with drift.*

Suppose now that $x_2 = \infty$. Then $I_g(x_2) = \infty$ and $I_f(x_2) < \infty$. We must now look at $I_h(x_2)$. We have

$$h(x) = f(x) \int_{x'}^x g(z) \, dz$$

$$= C \exp\left[-\frac{2\mu x}{\sigma^2} \right] \int_{x'}^x \exp\left[\frac{2\mu z}{\sigma^2} \right] dz$$

$$= k_1 \exp\left[-\frac{2\mu x}{\sigma^2} \right] \left(\exp\left[\frac{2\mu x}{\sigma^2} \right] - \exp\left[\frac{2\mu x'}{\sigma^2} \right] \right)$$

$$= k_1 + k_2 \exp\left[-\frac{2\mu x}{\sigma^2} \right]$$

where k_1 and k_2 are two additional constants. It can be seen that $I_k(x_2) = \infty$, so we may conclude that $x_2 = \infty$ is not an exit boundary. The possibility of an entrance boundary is excluded because $I_g(x_2) = \infty$. Hence $x_2 = \infty$ is a **natural boundary** for a Wiener process with drift. Similarly it is found that $x_1 = -\infty$ is also a natural boundary.

Transition densities for various boundary conditions

Example 1

Wiener process unrestricted on $(-\infty, \infty)$.

As just seen, the points $x_1 = -\infty$ and $x_2 = \infty$ are natural boundaries for the process being considered. The transition density must satisfy the Kolmogorov equation (12.14) and we have already found its solution in section 11.4. There we noted that $X(t)$ is a Gaussian random variable with mean $x + \mu t$ and variance $\sigma^2 t$. The required solution must therefore be

$$p(y, t | x) = \frac{1}{\sqrt{2\pi\sigma^2 t}} \exp\left[-\frac{(y - x - \mu t)^2}{2\sigma^2 t} \right]. \qquad (12.15)$$

This is called the *fundamental solution* or *Green's function* for the differential equation (12.14), because it is the result of applying an impulsive (delta-function) stimulus at the point x at time $t = 0$. The reader may verify that

$$\int_{-\infty}^{\infty} y p(y, t | x)\, dy = x + \mu t,$$

and that the corresponding variance is $\sigma^2 t$.

Example 2

Wiener process with an absorbing barrier at $x_2 = a < \infty$.

If $x_2 = a$ is finite then this is a regular boundary point and we may impose an absorbing barrier condition

$$p(a, t | x) = 0$$

on solutions of (12.14). We will find the required solution using the *method of images*.

   In the absence of the absorbing barrier at a, the solution is that of Example 1 - let us denote that solution by $p^\dagger(y, t | x)$. This, as we know, satisfies the Kolmogorov equation regardless of the starting value x. The Kolmogorov equation is linear in its subject p; hence any linear combination

$$c_1 p_1^\dagger + c_2 p_2^\dagger,$$

where c_1 and c_2 are arbitrary constants, will satisfy the same Kolmogorov equation if $p_1^\dagger$ and $p_2^\dagger$ do. We start a process off at the *image point*, $2a - x$, (that is the reflection of the start point in the absorbing barrier as mirror), to get $p^\dagger(y, t | 2a - x)$ and realize that the linear combination

$$p^\dagger(y, t | x) + c p^\dagger(y, t | 2a - x)$$

is also a solution. All we need do is to find the value of the constant c which makes the solution vanish at $y = a$. We require therefore that

$$0 = p^+(a, t|x) + cp^+(a, t|2a - x),$$

or, on substituting,

$$0 = \frac{1}{\sqrt{2\pi\sigma^2 t}}\left(\exp\left[-\frac{(a - x - \mu t)^2}{2\sigma^2 t}\right] + c\exp\left[-\frac{(x - a - \mu t)^2}{2\sigma^2 t}\right]\right).$$

This is rearranged to give

$$c = -\exp\left[\frac{-(a - x - \mu t)^2 + (x - a - \mu t)^2}{2\sigma^2 t}\right]$$

which simplifies to

$$c = -\exp\left[\frac{2\mu(a - x)}{\sigma^2}\right].$$

Hence the solution with an absorbing barrier at a is

$$p(y, t|x) = \frac{1}{\sqrt{2\pi\sigma^2 t}} \times$$

$$\left(\exp\left[-\frac{(y - x - \mu t)^2}{2\sigma^2 t}\right] - \exp\left[\frac{2\mu(a - x)}{\sigma^2}\right]\exp\left[-\frac{(y - 2a + x - \mu t)^2}{2\sigma^2 t}\right]\right).$$

$$(12.16)$$

To check that this satisfies the initial condition one only has to realize that the y-domain of the transition density is $(-\infty, a]$ and

$$\lim_{t \to 0} p^+(y, t|x) = \delta(y - x).$$

Example 3

Wiener process between reflecting barriers at 0 and a.

Note. This example contains a description of an exact method for solving the Kolmogorov equation. It may be omitted without creating a lack of continuity.

When a diffusion process is confined to a finite interval it is often possible to find the solution of the Kolmogorov equation as an infinite series by employing the method of *separation of variables*. This is a very useful technique, first employed by J. d'Alembert in 1749 for solving the wave equation, and since employed in many areas, including quantum mechanics, heat conduction etc. – in fact practically whenever a linear partial differential equation needs to be solved. It is the approach we will use here.

Accordingly, to consider a Wiener process with drift on $(0, a)$, with reflecting barriers at $y = 0$ and $y = a$, we solve the equation (12.14) with the initial condition

$$\lim_{t \to \infty} p(y, t \mid x) = \delta(y - x), x \in (0, a),$$

and with the boundary conditions of zero flux at $y = 0$ and $y = a$:

$$\left[-\mu p + \frac{\sigma^2}{2} \frac{\partial p}{\partial y} \right]_{y = 0, a} = 0.$$

To employ the separation of variables technique, we factor the transition density into a function of y and a function of t as

$$p(y, t \mid x) = T(t) Y(y),$$

the x-dependence being regarded as that of a parameter. Substituting TY for p in the Kolmogorov equation (12.14) and putting

$$D = \sigma^2$$

we get

$$T'Y = -\mu T Y' + D T Y''.$$

Dividing throughout by TY we obtain

$$\frac{T'}{T} = \frac{-\mu Y' + D Y''}{Y} = k.$$

Here k is called a *separation constant* – it is argued that if a function of t only is equal to a function of y only, then they must be mutually equal to a constant. To simplify we put

$$v = \frac{\mu}{D}, \quad \kappa = \frac{k}{D},$$

to obtain the two ordinary differential equations

$$T' = kT$$

$$Y'' - vY' - \kappa Y = 0.$$

The procedure now is to find values of κ (so-called **eigenvalues**) which lead to meaningful solutions, called **eigensolutions**.

First we put $\kappa = 0$ which implies that $T' = 0$ or $T = constant$. With this value of κ the differential equation for Y is

$$Y'' - vY' = 0,$$

with solutions of the form

$$Y(y) = c_1 + c_2 e^{vy},$$

where c_1 and c_2 are constants to be determined. Now the boundary conditions imply that

$$-Y + \frac{Y'}{v} = 0$$

and for this to be satisfied at $y = 0$ by $c_1 + c_2 e^{vy}$ we must have

$$c_1 = 0.$$

Applying the boundary condition at $y = a$ does not specifically yield c_2 which must therefore be found later by applying the normalization requirement $\int p \, dy = 1$. Hence, introducing the constant c_0, we set the zeroth eigensolution as

$$p_0(y, t) = c_0 e^{vy}.$$

Secondly we consider negative values of κ - there is no need to consider positive values because these would lead to solutions whose absolute values would increase exponentially indefinitely. We therefore put

$$\kappa = -\lambda^2.$$

so that

$$T(t) = \text{const.} e^{-\lambda^2 t}.$$

The differential equation for $Y(y)$ is now

$$Y'' - vY' + \lambda^2 Y = 0.$$

This is a second-order linear homogeneous differential equation with constant coefficients and is in standard form - see for example Berkey (1984) or other calculus texts. One tries a solution of the form

$$Y = e^{vy}$$

which on substituting in the differential equation yields

$$r^2 - vr + \lambda^2 = 0.$$

This is a quadratic equation for r with the two solutions

$$r_1 = \frac{v + \sqrt{v^2 - 4\lambda^2}}{2}$$

$$r_2 = \frac{v - \sqrt{v^2 - 4\lambda^2}}{2}.$$

The general form for $Y(y)$ is thus

$$Y(y) = c_1 e^{r_1 y} + c_2 e^{r_2 y}.$$

Determination of the non-zero eigenvalues

We must now apply the boundary conditions in order to find the particular values (**eigenvalues**) of r (or κ) which are allowed for the present problem. Now

$$Y'(y) = c_1 r_1 e^{r_1 y} + c_2 r_2 e^{r_2 y},$$

so the *boundary condition at* $y = 0$ gives

$$-c_1 - c_2 + \frac{c_1 r_1 + c_2 r_2}{v} = 0,$$

and the application of the *boundary condition at* $y = a$ gives

$$-c_1 e^{r_1 a} - c_2 e^{r_2 a} + \frac{c_1 r_1 e^{r_1 a} + c_2 r_2 e^{r_2 a}}{v} = 0.$$

These two equations may be written in vector notation as

$$\begin{pmatrix} \dfrac{r_1}{v} - 1 & \dfrac{r_2}{v} - 1 \\[2mm] \left(\dfrac{r_1}{v} - 1\right) e^{r_1 a} & \left(\dfrac{r_2}{v} - 1\right) e^{r_2 a} \end{pmatrix} \begin{pmatrix} c_1 \\ c_2 \end{pmatrix} = \begin{pmatrix} 0 \\ 0 \end{pmatrix}.$$

This system is in the form $\mathbf{Ac} = \mathbf{0}$ where $\mathbf{A}$ is a 2×2 matrix and $\mathbf{c}$ is a 2×1 column vector. From principles of elementary linear algebra (see for example Shields (1977) or Noble and Daniel (1978)), nontrivial (i.e., nonzero) solutions for c_1, c_2 can only be obtained if the determinant of $\mathbf{A}$ vanishes. Thus we require

$$\left(\frac{r_1}{v} - 1\right)^2 \left(\frac{r_2}{v} - 1\right)^2 [e^{r_2 a} - e^{r_1 a}] = 0.$$

Now because the possibilities that either $r_1 = v$ or $r_2 = v$ are excluded because they would imply that $\lambda = 0$, the case already considered, we are left with the requirement

$$e^{r_1 a} - e^{r_2 a} = 0.$$

On substituting for r_1 and r_2, we obtain, after factoring out $e^{v/2}$,

$$e^{va/2}[e^{a/2\sqrt{v^2 - 4\lambda^2}} - e^{-a/2\sqrt{v^2 - 4\lambda^2}}] = 0.$$

But because this is equivalent to the requirement

$$e^{a\sqrt{v^2 - 4\lambda^2}} = 1,$$

the possibility that $v^2 > 4\lambda^2$ is excluded (why?). Thus we must have

$$v^2 < 4\lambda^2$$

and may put for convenience

$$v^2 - 4\lambda^2 = -\rho^2.$$

Thus the eigenvalues are obtained from

$$e^{ia\rho/2} - e^{-ia\rho/2} = 0,$$

where $i = \sqrt{-1}$. Using Euler's formula $e^{i\theta} = \cos\theta + i\sin\theta$, this becomes the following simple equation for determining the eigenvalues,

$$\sin\left(\frac{a\rho}{2}\right) = 0.$$

Hence the eigenvalues are found from

$$\frac{a\rho}{2} = n\pi, n = 0, 1, 2, \ldots$$

or

$$\rho = \frac{2n\pi}{a},$$

there being no need to consider negative values of n becomes only ρ^2 enters the final expressions. Since $4\lambda^2 = v^2 + \rho^2$ we have

$$\lambda^2 = \frac{v^2}{4} + \frac{n^2\pi^2}{a^2},$$

so finally, substituting in the expressions for r_1 and r_2,

$$r_1 = \frac{v}{2} + i\frac{n\pi}{a},$$

and

$$r_2 = \frac{v}{2} - i\frac{n\pi}{a}.$$

The spatial (y-dependent) part of the n-th eigenfunction can therefore be expressed as the real-valued function

$$\phi_n(y) = e^{vy/2}\left[C_n\cos\left(\frac{n\pi y}{a}\right) + D_n\sin\left(\frac{n\pi y}{a}\right)\right],$$

where C_n and D_n are constants to be determined.

Determination of the constants C_n and D_n: One of the constants C_n, D_n can be eliminated by imposing the boundary condition $-Y + Y'/v = 0$ at $y = 0$.

This gives

$$
- e^{vy/2} \left[C_n \cos\left(\frac{n\pi y}{a}\right) + D_n \sin\left(\frac{n\pi y}{a}\right) \right]
$$

$$
+ \frac{1}{v} e^{vy/2} \left[\frac{v}{2} \left\{ C_n \cos\left(\frac{n\pi y}{a}\right) + D_n \sin\left(\frac{n\pi y}{a}\right) \right\} \right.
$$

$$
\left. - \frac{n\pi}{a} \left\{ C_n \sin\left(\frac{n\pi y}{a}\right) - D_n \cos\left(\frac{n\pi y}{a}\right) \right\} \right]\Bigg|_{y=0} = 0.
$$

This simplifies to

$$
- C_n + \frac{1}{v} \left[\frac{v}{2} C_n - \frac{n\pi}{a} D_n \right] = 0
$$

or

$$
D_n = - \frac{av}{2n\pi} C_n.
$$

Now we utilize the fact that for any linear differential equation, a linear combination of an arbitrary number of solutions also satisfies the equation. Thus we may form the general solution of the forward Kolmogorov (Fokker–Planck) equation for the transition probability density function for a Wiener process with drift between two reflecting boundaries as the infinite sum

$$
p(y,t\,|\,x) = C_0 e^{vy} + e^{vy/2} \sum_{n=1}^{\infty} C_n e^{-\{v^2/4 + n^2\pi^2/a^2\}t} \left[\cos\left(\frac{n\pi y}{a}\right) - \frac{va}{2n\pi} \sin\left(\frac{n\pi y}{a}\right) \right].
$$

Let us define the *spatial eigenfunctions*. Firstly the zeroth order eigenfunction,

$$
Y_0(y) = e^{vy}
$$

and for $n = 1, 2, \ldots$

$$
Y_n(y) = e^{vy/2} \left[\cos\left(\frac{n\pi y}{a}\right) - \frac{va}{2n\pi} \sin\left(\frac{n\pi y}{a}\right) \right].
$$

Then it will be shown in the exercises that

$$
\int_0^a Y_m(y) Y_n(y) e^{-vy} \, dy = A_n \delta_{mn}, \tag{12.17}
$$

where δ_{mn} is *Kronecker's delta* defined by

$$
\delta_{mn} = \begin{cases} 1, & \text{if } m = n; \\ 0, & \text{otherwise,} \end{cases}
$$

and the constants A_n, which will be found below, are called *normalization constants.* When functions satisfy a relation of the kind (12.17) they are said to be *orthogonal with a weight function of* e^{-vy}.

Let us show how (12.17) is used to find the constant coefficients C_n. We have as the initial condition

$$p(y, 0 | x) = \sum_{m=0}^{\infty} C_m Y_m(y) = \delta(y - x).$$

Now multiply this by $Y_n(y)e^{-vy}$ and integrate from 0 to a:

$$\int_0^a \sum_{m=0}^{\infty} C_m Y_m(y) Y_n(y) e^{-vy} \, dy = \int_0^a Y_n(y) e^{-vy} \delta(y - x) \, dy.$$

Hence on utilizing the substitution property of the delta-function,

$$C_n = C_n(x) = \frac{Y_n(x) e^{-vx}}{A_n}.$$

We have therefore as the final expression for the transition probability density function of a Wiener process with drift parameter μ and variance parameter σ^2 confined between reflecting barriers at 0 and a, starting at $x \in (0, a)$,

$$p(y, t | x) = C_0 e^{vy} + e^{-v^2 t/4} \sum_{n=1}^{\infty} \frac{Y_n(x) Y_n(y) e^{-vx}}{A_n} e^{-(n^2 \pi^2 / a^2)t}, \qquad (12.18)$$

where $v = \mu/\sigma^2$ and the constants C_0 $\{A_n, n = 1, 2, \ldots\}$ are given in the exercises.

Limiting behaviour at $t = \infty$ – stationary distributions

Let us examine the ultimate ($t = \infty$) behaviour of the probability distribution of $X(t)$ for the Wiener process with drift in the above three cases.

Example 1

For the Wiener process with drift unrestricted by barriers on the whole real line a stationary distribution does not exist. This may be seen by using the formulae of section 12.2. For a process on an interval whose end-points are natural boundaries, the stationary density is

$$p^*(y) = \frac{K_2}{\beta(y) \phi(y)}.$$

Since $\alpha(y) = \mu$ and $\beta(y) = \sigma^2$, we have

$$\phi(y) = \exp\left[2 \int^y \frac{\mu}{\sigma^2} \, dz \right]$$

$$= e^{2\mu y/\sigma^2}.$$

Thus the constant K_2 is

$$K_2 = 1 \Big/ \left(\int_{-\infty}^{\infty} [\sigma^2 e^{2\mu y/\sigma^2}]^{-1} \, dy \right)$$

$$= 0.$$

Thus the distribution is *degenerate*.

This can also be seen by examining the limiting behaviour of $p(y, x|t)$ as $t \to \infty$. It will be seen that

$$\lim_{t \to \infty} p(y, t|x) = 0,$$

for all $y \in (-\infty, \infty)$.

Example 2

For a Wiener process with drift on $(-\infty, a)$, with an absorbing barrier at a, the transition density is given by (12.16) for $y < a, t < \infty$. It can be shown by the reader that this $p(y, t|x) \to 0$ as $t \to \infty$ for all $y \in (\infty, a)$. Again, therefore, the stationary distribution is degenerate.

A little more can be said, however. Suppose we define $p_A(t; a, x)$ as the probability that the process has been absorbed at or before time t. Then the probability is $1 - p_A(t; a, x)$ that the process has not reached the absorbing barrier a at time t, and is thus still 'wandering around' below the barrier. This is given by

$$1 - p_A(t; a, x) = \Pr\{X(t) < a | X(0) = x < a\}$$

$$= \int_{\infty}^{a} p(y, t|x) \, dy.$$

When the expression (12.16) is inserted for $p(y, t|x)$, one obtains

$$1 - p_A(t; a, x) = \Phi\left(\frac{a - x - \mu t}{\sqrt{\sigma^2 t}} \right) - e^{2u(a - x)/\sigma^2} \Phi\left(\frac{x - a - \mu t}{\sqrt{\sigma^2 t}} \right) \qquad (12.19)$$

where $\Phi(.)$ is the standard *normal distribution function* defined by

$$\Phi(z) = \int_{-\infty}^{z} \frac{e^{-y^2/2}}{\sqrt{2\pi}} \, dy,$$

and for which tables are given in most introductory statistics texts. We have

$$p_A(t; a, x) = \Pr\{X(t) = a | X(0) = x < a\}$$

and as already implied

$$\Pr\{X(t) \in (-\infty, a] | X(0) = x < a\} = 1,$$

where this latter quantity is the probability that the process is in the union of the absorbing point-set $\{a\}$ and the free part of the domain, the semi-infinite open interval $(-\infty, a)$.

In Exercise 7 it will be seen that the behaviour of p_A as $t \to \infty$ depends on μ. If $\mu \geq 0$, so there is either zero drift or there is drift towards the barrier, then

$$p_A(t; a, x) \to 1.$$

Thus in these cases, **absorbtion at any finite barrier point is eventually certain.**

In the remaining cases, when $\mu < 0$ so that the drift is away from the barrier, the probability of ultimate absorbtion is less than one and is given by

$$p_A(t; a, x) \to e^{2\mu(a-x)/\sigma^2}.$$

Example 3

For a Wiener process confined between two reflecting barriers, a non-degenerate stationary distribution exists. This is because the initial probability mass cannot escape nor be absorbed at the boundaries. The corresponding density can be found either by the method outlined in section 12.2 or by considering the behaviour of $p(y, t|x)$ given by (12.18) as $t \to \infty$. We will leave the first approach as an exercise. For the second approach, note that for large values of t, the transition probability density is well-approximated by taking the first two terms in the eigenfunction expansion for p:

$$p(y, t|x) \sim C_0 e^{vy} + \frac{Y_1(x) Y_1(y)}{A_1} \exp\left[-\left(\frac{v^2}{4} + \frac{\pi^2}{a^2} \right) t \right]$$

$$\to C_0 e^{vy}$$

as $t \to \infty$. This shows that the stationary density $p^*(y)$ is of the form $C_0 e^{vy}$. The constant is determined by the requirement

$$\int_0^a p^*(y)\, dy = 1,$$

so that

$$C_0 = 1 \bigg/ \int_0^a e^{vy}\, dy$$

$$= \frac{v}{e^{va} - 1}.$$

Thus the **stationary density** is

$$p^*(y) = \frac{v e^{vy}}{e^{va} - 1}.$$

Note that one may proceed without necessarily following the procedure exactly as outlined in section 12.2. One just assumes that since the required solution of the forward Kolmogorov equation is time independent, the stationary density must satisfy the ordinary differential equation

$$-\mu \frac{\mathrm{d}p^*}{\mathrm{d}y} + \frac{\sigma^2}{2} \frac{\mathrm{d}^2 p^*}{\mathrm{d}y^2} = 0,$$

with boundary conditions

$$-\mu p^* + \frac{\sigma^2}{2} \frac{\mathrm{d}p^*}{\mathrm{d}y}\bigg|_{y=0, y=a} = 0.$$

On solving this equation and applying the normalization condition, one will obtain the appropriate expression for $p^*(y)$.

12.4 THE ORNSTEIN-UHLENBECK PROCESS

Whereas the Wiener process can be viewed as a mathematical representation for the position (or displacement) of a particle undergoing Brownian motion, as in section 7.8, or as an approximation for a random walk, the *Ornstein-Uhlenbeck process* (**OUP**) was introduced as a mathematical model for the velocity of a Brownian motion particle (Uhlenbeck and Ornstein, 1930). These authors employed the following stochastic differential equation for the velocity $v(t)$ of the particle at time t:

$$\frac{\mathrm{d}v}{\mathrm{d}t} + kv = w(t),$$

where k is constant and w is a white noise. Unlike the Wiener process, the OUP unrestricted on $(-\infty, \infty)$ has a non-degenerate stationary distribution. Physically, since $\mathrm{d}v/\mathrm{d}t$ is acceleration, this can be traced to the force which always acts to pull the process (particle) back to the origin, the force being correspondingly larger as the distance from the origin increases.

In addition to its initial use in studies of Brownian motion, the OUP has found applications in diverse areas such as, *inter alia*, astrophysics (Chandrasekhar, 1943) stochastic models for neuronal activity (Ricciardi, 1977; Tuckwell, 1988, 1994), birth and death theory (McNeil and Schach, 1973), population biology (Tuckwell, 1974) and models for noisy currents through ionic channels in membranes (Tuckwell, 1987).

Infinitesimal moments and Kolmogorov equations

We may define the OUP as a time-homogeneous diffusion process with
(i) first infinitesimal moment proportional to the value of the process, so

$$\alpha(x, t) = -ax,$$

where a is a positive constant. We say the **drift** is directed towards the origin. A constant drift sometimes occurs naturally in addition to the drift proportional to x and this does not affect the overall qualitative properties of the process. Denoting this constant by b, which can be any real number, the more general form of the first infinitesimal moment is

$$\alpha(x, t) = -ax + b;$$

(ii) a constant second infinitesimal moment, so

$$\beta(x, t) = \sigma^2,$$

where σ is a non-negative real constant.

The forward Kolmogorov equation satisfied by the transition density $p(y, t|x)$ is, from (12.4),

$$\frac{\partial p}{\partial t} = \frac{\partial}{\partial y}((ay - b)p) + \frac{\sigma^2}{2}\frac{\partial^2 p}{\partial y^2},$$

and the corresponding backward equation is

$$\frac{\partial p}{\partial s} = (b - ax)\frac{\partial p}{\partial x} + \frac{\sigma^2}{2}\frac{\partial^2 p}{\partial x^2}.$$

Boundary classification

The calculations of the integrals, defined in section 12.1, which are required in order to ascertain the natures of various boundary points for an OUP, are more complicated than those for the Wiener process. It is left as an exercise to show that for an OUP:

  (i) **infinite boundary points**, $-\infty$ and ∞, are natural boundaries, and
  (ii) any **finite boundary point** $c\in(-\infty, \infty)$ is a regular boundary.

One sees, therefore, that boundary points have the same natures for the OUP as they did for a Wiener process with drift. The processes, however, have very different properties. One example is the existence of a non-degenerate stationary distribution for the unrestricted OUP but not for a Wiener process; another example is that any finite point is eventually reached with probability one in a finite time by any OUP, regardless of the values of the parameters a, b and σ, whereas this is not true for a Wiener process - the sign of the drift is important - in some cases one may wait indefinitely and not be assured that certain values are reached.

Transition densities for various boundary conditions

Example 1

OUP unrestricted on $(-\infty, \infty)$

Even though the forward Kolmogorov equation satisfied by $p(y, t|x)$ appears fairly simple, it is by no means an easy partial differential equation to solve exactly. We need to find a solution, bounded on $(-\infty, \infty)$ with initial value $\delta(y-x)$.

We will employ a fact, to be proved in Exercise 12.27, which will enable us to determine the transition density immediately. We utilize a simple relation between the OUP and a Wiener process with drift. The relation enables us to state that if the initial value of an OUP is x, then at time t, the random variable $X(t)$ i.e., the value of an OUP at time t, is a **Gaussian random variable** with mean

$$E[X(t)| X(0) = x] = xe^{-at} + \frac{b}{a}(1 - e^{-at}),$$

and variance

$$\text{Var}[X(t)| X(0) = x] = \frac{\sigma^2}{2a}(1 - e^{-2at}).$$

Hence, on using the standard formula for a normal density, we must have

$$p(y, t \mid x) = \left(\frac{a}{\pi\sigma^2(1 - e^{-2at})} \right)^{1/2}$$

$$\times \exp \left[\frac{-\left[y - xe^{-at} - \frac{b}{a}(1 - e^{-at}) \right]^2}{\frac{\sigma^2}{a}(1 - e^{-2at})} \right].$$

We have been able to solve a Kolmogorov differential equation, practically by inspection, using a probabilistic argument. This illustrates the interesting procedure whereby a connection is made between two different branches of mathematics - in this case, probability and partial differential equations.

Example 2

The OUP on other intervals

The transition pdfs on an OUP on semi-infinite and finite intervals can be found for various boundary conditions using the method of separation of variables. Finding these solutions involves a large amount of labour. The interested student is left to find such solutions as an exercise.

Limiting behaviour at $t = \infty$ – stationary distributions

As we have mentioned at the beginning of this section, the OUP on $(-\infty, \infty)$ differs markedly from the Wiener process in that a stationary density exists for all parameter values. This time-independent density can be found quickly by noting that the mean and variance of $X(t)$ take the following values as

$t \to \infty$, regardless of the initial value $X(0)$:

$$E[X(t)] \to \frac{b}{a},$$

$$\mathrm{Var}[X(t)] \to \frac{\sigma^2}{2a}.$$

The stationary density is that of a Gaussian random variable with these values for its mean and variance and is thus

$$p^*(y) = \frac{1}{\sqrt{\pi\sigma^2/a}} \exp\left[\frac{-\left(y - \frac{b}{a}\right)^2}{\sigma^2/a}\right].$$

The approach of $p(y, t \mid x)$ to the stationary density $p^*(y)$ is illustrated in Fig. 12.1.

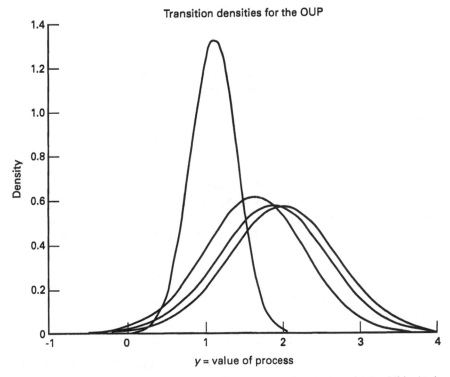

Figure 12.1 Showing the approach of the transition density of an Ornstein-Uhlenbeck process to its stationary form. The parameter values are $a = 1$, $b = 1$, $\sigma = 1$ and the initial value is $x = 0$. The transition density is shown for $t = 0.1$, $t = 1$, $t = 2$ and $t = 5$. Note that for all t the density is Gaussian. The peak moves from left to right and by $t = 5$ the density is practically indistinguishable from the stationary form.

Alternatively, the formula (see section 12.2)

$$p^*(y) = \frac{K_2}{\beta(y)\phi(y)},$$

with

$$\phi(y) = \exp\left[\int^y \frac{2(b - ay')}{\sigma^2} \, dy'\right].$$

can be employed – see the exercises.

12.5 STOCHASTIC INTEGRALS AND STOCHASTIC DIFFERENTIAL EQUATIONS

We have seen in section 11.4 that a Wiener process with drift can be characterized by the *stochastic differential equation*

$$dX = \mu \, dt + \sigma \, dW.$$

The correct interpretation of this equation is in terms of an integral involving a Wiener process – called a **stochastic integral**.

There are a large number of integrals which one may define in connection with random processes. Mathematical complexities arise when integrals involving W are considered because of the irregular properties of the paths of W. This means that the methods of defining integrals given in real-variable calculus courses cannot be used. We will consider stochastic integrals very briefly and somewhat superficially – there are numerous technical accounts – see for example Gihman and Skorohod (1972), Arnold (1974), Lipster and Shiryayev (1977) or Oksendal (1985). Our main purpose is to enable the reader to understand and know how to use a stochastic differential equation of the general form

$$dX(t) = f(X(t), t) \, dt + g(X(t), t) \, dW(t).$$

Equivalently, dropping the reference to t in the random processes, we can write this as

$$dX = f(X, t) \, dt + g(X, t) \, dW, \tag{12.20}$$

where f and g are real-valued functions, W is a standard Wiener process and X is a random process which in cases of interest will be a diffusion process.

However, it must be stated at the outset that (12.20) does **not** always have a unique interpretation. This situation arises for the following reason. Equation (12.20) is interpreted correctly as implying the **stochastic integral**

equation

$$X(t) = X(0) + \int_0^t f(X(t'), t') \, dt' + \int_0^t g(X(t'), t') \, dW(t'), \qquad (12.21)$$

and the process X so defined is called a **solution of the stochastic differential equation (12.20)**. Although the first integral here presents no problems, there are many ways of defining the second one,

$$\int_0^t g(X(t'), t') \, dW(t'),$$

which is called a **stochastic integral**. Furthermore, the different definitions can lead to various solutions, X, with quite different properties.

Despite this apparent ambiguity, there are two useful definitions which are most commonly employed – the **Ito stochastic integral** and the **Stratonovich stochastic integral**; and there is a simple relation between these two.

A note on notation. It is preferable in (12.20) not to 'divide' throughout by dt, because as we have seen, the derivatives of W and hence of X do not exist in the usual sense. However, as long as we keep that in mind, it is possible to display (12.20) as a stochastic differential equation involving *white noise w*, the 'derivative' with respect to time t of W (see section 11.3):

$$\frac{dX}{dt} = f(X, t) + g(X, t)w,$$

or perhaps even

$$\frac{dX}{dt} = f(X, t) + g(X, t)\dot{W}.$$

Stochastic differential equations written in this form are often called **Langevin equations**.

Let us now make an important observation on the stochastic differential equation (12.20). **If the function g is identically zero, the differential equation is deterministic** and can be written in the usual way

$$\frac{dX}{dt} = f(X, t).$$

Assuming the initial value $X(0) = x_0$ is not random then $X(t)$ is non-random for all t and this equation is solved in the usual way.

We expect that the behaviour of solutions of this deterministic equation would be related to those of the stochastic differential equation (12.20), and be close to them when the **noise term** g is small. We would be correct in believing that, in particular, the expected value $E[X(t)]$ of the solution of

(12.20) would not be very far, in most cases, from the solution of the deterministic equation.

This can be illustrated nicely with the Wiener process with linear drift μt and variance parameter σ. From section 11.4, this process has the stochastic differential equation (11.11):

$$dX = \mu \, dt + \sigma \, dW.$$

Here

$$f(X, t) = \mu,$$

and

$$g(X, t) = \sigma.$$

If we put $\sigma = 0$ we obtain the deterministic differential equation

$$\frac{dX}{dt} = \mu.$$

The solution of this with initial value x_0 is

$$X(t) = x_0 + \mu t,$$

and this, as seen in section 11.4, is equal to the mean value function of the process satisfying (11.11). The added noise makes the paths of X very irregular, but the mean value is still μt. This was depicted in Fig. 11.4.

Heuristic interpretation

Before proceeding more formally, let us describe roughly how we can understand an equation of the form (12.20). This can perhaps best be accomplished by writing the related difference equation

$$\Delta X = f(X, t)\Delta t + g(X, t)\Delta W. \tag{12.22}$$

Here we may regard the (random) *increment* in X in the time interval $(t, t + \Delta t]$ as having two components. The first component is equal to the value of $f(X, t)$ at the beginning of the time interval multiplied by the length Δt of the time interval. The second component is the value of $g(X, t)$ at the beginning of the time interval, multiplied by the (random) increment $\Delta W = W(t + \Delta t) - W(t)$ that occurs in a standard Wiener process in Δt. As we have seen, ΔW is a Gaussian random variable with mean zero and variance Δt. We have essentially outlined a **method of simulation** of the stochastic differential equation (12.20) – this will be elaborated on below.

It should be realized, however, that even though the functions f and g are functions in the usual deterministic sense, both the components of the increment in X, namely, $f\Delta t$ and $g\Delta W$, are random variables, because

$f(X(t), t)$, as well as $g(X(t), t)$ are random variables by virtue of their dependence on the value of the random variable $X(t)$.

We will now proceed to define the Ito integral, then state, with the aid of a proof outline, a change of variable rule called **Ito's formula**. This will be followed by a brief consideration of the Stratonovich integral and the roles of the various integrals in stochastic modelling.

The Ito stochastic integral

We are going to define the Ito stochastic integral

$$I = \int_0^b f(t)\,dW(t),$$

where $\{f(t), t \in [a, b)\}$ is a suitable random process. (The f here is general and not related to that in (12.20).) Suitable random processes will be said to belong to **class** M. We define this class as containing random processes f which satisfy the following requirements:

(i) f is a *non-anticipative* process. In the present context this means that questions about f up to and including time t can be answered without knowledge of the evolution of the Wiener process W for times beyond t; or one could say that the evolution of f up to and including any particular time is *independent* of future values of W.

(ii) the integral of the square of $f(t)$ over $[a, b)$ is *finite* with probability one; that is,

$$\Pr\left\{ \int_a^b [f(t)]^2\,dt < \infty \right\} = 1.$$

Thus the sample paths of f cannot be often singular or too wildy fluctuating – such paths may occur, but their associated probability is zero. When an event occurs with probability one it is said to be **almost sure** so we could say that the integral of f^2 is almost surely finite. Note that the integral appearing here is, like the Ito integral we are about to define, a random variable, which takes on various values as the various sample paths of f arise.

Definition of Ito stochastic integral for simple random functions

A **simple random function** (or step function), in the present context, is one which is constant on sub-intervals of $[a, b)$. Such a function is represented *schematically* in Fig. 12.2. The constant values on the various sub-intervals are actually *random variables* – fixed ones – rather than numbers as in real-variable calculus. We will insert the underlying sample-space variable ω to make it clear that these are constant random variables. Thus we may put

$$f(t, \omega) = f_j(\omega), \quad t_j \leqslant t < t_{j+1}, \quad j = 0, 1, \ldots, n - 1,$$

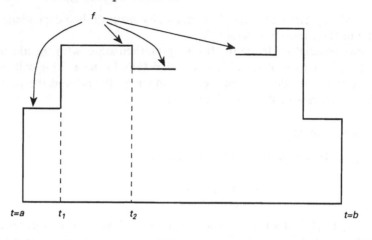

Figure 12.2 A *schematic* representation of a simple function f. In the present context the constant values on sub-intervals are actually random variables.

where $a = t_0 < t_1 < \cdots < t_{n-1} < t_n = b$ is a **partition** of the interval $[a, b)$ and the $\{f_j\}$ are a set of n random variables.

Now for each sub-interval $[t_j, t_{j+1})$ of $[a, b)$ we can define the corresponding increment in a standard Wiener process

$$\Delta W_j = W(t_{j+1}) - W(t_j), \quad j = 0, 1, \ldots, n - 1,$$

which defines a further n random variables. We are now ready to define the Ito stochastic integral of a simple random function f with respect to a Wiener process as

$$\int_a^b f(t)\, dW(t) = \sum_{j=0}^{n-1} f_j \Delta W_j.$$

We note that we could also write this as

$$\int_a^b f(t)\, dW(t) = \sum_{j=0}^{n-1} f(t_j)[W(t_{j+1}) - W(t_j)] \tag{12.23}$$

which will be a useful observation when we distinguish between the Ito and Stratonovich integrals.

In order to define the stochastic integral for the general class of random processes we have called M, we state the following lemma without proof - see the references at the beginning of this section. This lemma tells us that for any random process f in M, we can be sure there is a sequence $\{f_n, n = 1, 2, \ldots\}$, also in M, whose members get closer and closer to f as $n \to \infty$, in the following sense.

Lemma   **Let $f \in M$ be a random process in the class defined above. Then a sequence of simple functions $f_n(t) \in M$ exists such that as $n \to \infty$,**

$$\int_a^b [f(t) - f_n(t)]^2 \, dt \xrightarrow{\;p\;} 0.$$

That is, this sequence of integrals, which measure the distances between f and f_n, converges in probability to zero – recall the definition of this mode of convergence in section 6.6.

Definition of Ito stochastic integral for arbitrary $f \in M$

Since we know how to define the Ito stochastic integral for simple functions, we extend the definition to arbitrary random processes in M by using a sequence of approximating integrals of simple functions. The limit of this sequence is defined as the required integral. Thus we set

$$\int_a^b f(t) \, dW(t) = \lim_{n \to \infty} \int_a^b f_n(t) \, dW(t),$$

where the limit is again in the sense of **convergence in probability of a sequence of random variables**. The above lemma guarantees that a suitable sequence $\{f_n(t)\}$ of approximating random functions exists.

We mentioned that other definitions can be given for the Ito stochastic integral according to the different properties which are ascribed to the integrand $f(t)$. Usually these conditions are more restrictive (**stronger**) than the ones we have employed. However, then, and indeed in most cases of practical interest, we have the following results concerning the mean and the variance of the Ito stochastic integral:

(i) Mean

$$E\left(\int_a^b f(t) \, dW(t) \right) = 0. \tag{12.24}$$

(ii) Variance

$$E\left(\int_a^b f(t) \, dW(t) \right)^2 = \int_a^b f^2(t) \, dt. \tag{12.25}$$

We will not prove (12.25) but (12.24) will be considered in the exercises. Furthermore, if f and g are both in M then

$$E\left(\int_a^b f(t) \, dW(t) \times \int_a^b g(t) \, dW(t) \right) = \int_a^b E[f(t)g(t)] \, dt. \tag{12.25'}$$

When the Ito definition of stochastic integral is used in the interpretation of equation (12.20) we call it an *Ito stochastic differential* or *Ito stochastic differential equation*. The process X is called the solution of the equation. There are many accounts of existence and uniqueness theorems and properties of the solution – see for example the references at the beginning of this section. We will avoid technicalities and simply assume that the functions f and g are *suitable*.

Ito's formula

When one uses Ito's definition of stochastic integral, the usual rules of calculus sometimes must be modified. This is not the case with the Stratonovich definition considered below. In particular, one must be careful when changing variables as the following result, called **Ito's formula**, shows.

We will not give a detailed proof, which can be found for example in Liptser and Shiryayev (1977). We will simply utilize the fact that

$$(dW^2) \text{ acts like } dt. \tag{12.26}$$

This statement can be understood in terms of the following probability one relation (see, for example the previous reference) for the so-called *quadratic variation* of W over the interval $[0, t)$:

$$\lim_{n \to \infty} \sum_{j=0}^{n-1} [W(t_{j+1}) - W(t_j)]^2 = t,$$

where $\{0 = t_0 < t_1 < \cdots < t_n = t\}$ is a partition of $[0, t)$. This suggests we can write

$$\int_0^t (dW)^2 = t.$$

and the latter can be written as (12.26). The significance of this result for our purposes is that whereas second order differentials such as $(dt)^2$ can usually be ignored, $(dW)^2$ is of the same order as dt and hence cannot be neglected.

We are now ready to discuss Ito's formula.

Notation. In the following we will use subscripts to denote partial differentiation with respect to a variable. Thus, for example, if $h = h(x, t)$ then

$$h_{xt} = \frac{\partial^2 h}{\partial t \partial x}.$$

Theorem 12.4 **Let the random process X have the Ito stochastic differential**

$$dX = f(X, t) \, dt + g(X, t) \, dW, \tag{12.27}$$

for suitable functions f and g. Suppose we change variables by putting

$$Y = h(X, t).$$

Then Y satisfies the Ito stochastic differential equation

$$\mathrm{d}Y = \left[fh_x + \frac{g^2}{2} h_{xx} + h_t \right] \mathrm{d}t + h_x g \, \mathrm{d}W, \tag{12.28}$$

where h_x and h_t are the first partial derivatives of the function h with respect to its first and second variables, respectively, and h_{xx} is the second partial derivative of h with respect to its first variable. Equation (12.28) is called Ito's formula.

Proof outline It we use the first few terms in a Taylor's expansion for h we get

$$\begin{aligned}
\mathrm{d}Y &= h_x(X,t)\,\mathrm{d}X + h_t(X,t)\,\mathrm{d}t \\
&\quad + \tfrac{1}{2}h_{xx}(X,t)(\mathrm{d}X)^2 + h_{xt}(X,t)\,\mathrm{d}X\,\mathrm{d}t + \tfrac{1}{2}h_{tt}(\mathrm{d}t)^2 + \cdots \\
&= h_x(X,t)[f\,\mathrm{d}t + g\,\mathrm{d}W] + h_t(X,t)\,\mathrm{d}t \\
&\quad + \tfrac{1}{2}h_{xx}(X,t)[f^2(\mathrm{d}t)^2 + 2fg\,\mathrm{d}t\,\mathrm{d}W + g^2(\mathrm{d}W)^2] \\
&\quad + h_{xt}(X,t)[f\,\mathrm{d}t + g\,\mathrm{d}W]\,\mathrm{d}t + \tfrac{1}{2}h_{tt}(X,t)(\mathrm{d}t)^2 + \cdots
\end{aligned}$$

Retaining only those terms of order $\mathrm{d}t$ or $\mathrm{d}W$, we obtain (12.28). There is the extra term involving h_{xx} which is absent in standard calculus.

Example

Suppose X is a standard Wiener process and so satisfies the stochastic differential equation

$$\mathrm{d}X = \mathrm{d}W.$$

What Ito stochastic equation does the new variable

$$Y = X^2$$

satisfy?

Solution

Relating this to the standard form (12.27), we find $f(X,t) = 0$ and $g(X,t) = 1$. Now $h(X,t) = X^2$, so we have

$$\begin{aligned}
h_x(X,t) &= 2X \\
h_{xx}(X,t) &= 2 \\
h_t(X,t) &= 0.
\end{aligned}$$

Using (12.28) we see

$$dY = [2X.0 + \tfrac{1}{2}.2 + 0]\,dt + 2X\,dW$$

or

$$dY = 1\,dt + 2\sqrt{Y}\,dW,$$

which is the required stochastic differential equation for Y. The extra term is $1\,dt$. Other examples will appear as exercises.

Stratonovich's stochastic integral

Consider equation (12.23) where the Ito integral was defined for simple functions. The Stratonovich integral differs in that instead of employing the value of f at the beginning t_j of the sub-interval $[t_j, t_{j+1})$, the value at the mid-point $(t_j + t_{j+1})/2$ is used. The integral is thus said to be *symmetrized*. This difference is sufficient to alter sometimes the properties of the resulting random variable significantly.

With the Stratonovich definition, the change of variable formula (12.28) becomes simply

$$dY = [fh_x + h_t]\,dt + h_x g\,dW, \tag{12.29}$$

and usual calculus rules (product rule, quotient rule, etc.) can be used.

However, in most cases encountered in modelling real-world phenomena, one may switch back and forth from the Ito and Stratonovich schemes by using the following simple result. If X satisfies the *Stratonovich stochastic differential equation*

$$dX = f(X, t)\,dt + g(X, t)\,dW,$$

then the corresponding Ito equation is

$$dX = [f(X, t) + \tfrac{1}{2}g(X, t)g_x(X, t)]\,dt + g(X, t)\,dW. \tag{12.30}$$

Note that if g contains no explicit X-dependence, then g_x is zero and there is no difference between the two definitions. In particular, **for all equations of the form**

$$dX = f(X, t)\,dt + g(t)\,dW,$$

it makes absolutely no difference whether the interpretation is through an Ito or a Stratonovich stochastic integral.

Example

If X satisfies the Stratonovich equation

$$dX = X\,dt + X^2\,dW,$$

what is the equivalent Ito equation?

Solution

Here $f = X$ and $g = X^2$. Hence $g_x = 2X$ and we find

$$dX = [X + \tfrac{1}{2}X^2.2X]\,dt + X^2\,dW$$

or

$$dX = [X + X^3]\,dt + X^2\,dW.$$

is the corresponding Ito equation.

12.6 MODELLING WITH STOCHASTIC DIFFERENTIAL EQUATIONS

We have seen in the last section that a wide class of random processes may be defined directly by writing down a stochastic differential equation of the form

$$dX = f(X, t)\,dt + g(X, t)\,dW,$$

where we call $f(X, t)\,dt$ the **drift term** and $g(X, t)\,dW$ the **noise term**. In most cases of interest, the solutions of such equations are *diffusion processes* which were defined by analytical methods in section 12.1. In that section, diffusion processes were defined in terms of their *first infinitesimal moment* - given as $\alpha(x, t)$ in Equation (12.1) - and *infinitesimal variance* - given as $\beta(x, t)$ in Equation (12.2). Such processes have sample paths which are **continuous** with probability one.

It would be very convenient to have a relationship between these two representations so that given a stochastic differential equation for X one could immediately ascertain its infinitesimal mean and variance, and *vice versa*: - given the infinitesimal moments, write down the corresponding stochastic differential equation. It is indeed possible to accomplish these switches very easily as the following result shows.

Theorem 12.5 **Let X be a diffusion process with Ito stochastic differential equation**

$$dX = f(X, t)\,dt + g(X, t)\,dW,$$

where f and g are suitable functions. Then the infinitesimal mean of X is given by

$$\alpha(x, t) = f(x, t),$$

and the infinitesimal variance of X is given by

$$\beta(x, t) = g^2(x, t).$$

Thus the differential equations (12.4) and (12.5) satisfied by the transition probability density function can be obtained simply from the stochastic differential equation.

In particular, for a **time-homogeneous** process with stochastic equation

$$dX = f(X)\,dt + g(X)\,dW,$$

the forward Kolmogorov equation for the transition density $p(y, t\,|\,x)$ is, from Equation (12.6),

$$\frac{\partial p}{\partial t} = - \frac{\partial (f(y)p)}{\partial y} + \frac{1}{2}\frac{\partial^2(g^2(y)p)}{\partial y^2},$$

and the corresponding backward equation is, from Equation (12.7),

$$\frac{\partial p}{\partial t} = f(x)\frac{\partial p}{\partial x} + \frac{1}{2}g^2(x)\frac{\partial^2 p}{\partial x^2}.$$

If a stochastic differential equation is given with Stratonovich's interpretation, it may first be converted to an Ito equation using Equation (12.30); then Theorem 12.5 will yield the infinitesimal moments of the process.

It might seem that, in modelling a random phenomenon with a stochastic differential equation, an ambiguity arises in the choice of an Ito or Stratonovich interpretation. This, however, is not the case. A diffusion process is specified once its infinitesimal mean and variance are given. Thus, when deriving a model one must be certain that these infinitesimal moments are correct. Once this is done it matters not which stochastic calculus one adopts.

A word of caution is necessary. It is not a good idea to take an existing deterministic differential equation and convert one of its parameters to white noise. The reason is that in most cases there **will be ambiguity** because there is no *a priori* reason why a particular interpretation, Ito or Stratonovich, should be correct. A particular choice would only be defensible if the infinitesimal moments could be ascertained to be the correct ones. This will be illustrated in the exercises. However, in certain cases it has been shown that limits of sequences of discrete stochastic equations have solutions corresponding to the Stratonovich differential (Wong and Zakai, 1965).

12.7 APPLICATIONS

Diffusion approximation to a random walk

Suppose a random walk $X_\varepsilon = \{X_\varepsilon(t), t \geq 0\}$ occurs with jumps up or down of magnitude ε. Jumps up form a Poisson process N_1 with intensity λ and jumps down form another Poisson process N_2, independent of N_1, but with the same intensity. One may assume for convenience that the processes start at zero at time $t = 0$. Note that this random walk, sometimes called a **randomized random walk** (Feller, 1971), differs from the simple random walk of Chapter 7 in that there the jumps (steps) occurred at fixed time intervals whereas now they occur at random times.

We may write the following expression for the value of the process X_ε at time t:

$$X_\varepsilon(t) = \varepsilon N_1(t) - \varepsilon N_2(t).$$

Similarly we may write the following expression for the increment in X_ε in the interval $(t, t + \Delta t]$:

$$\begin{aligned}
\Delta X_\varepsilon &= X_\varepsilon(t + \Delta t) - X_\varepsilon(t) \\
&= \varepsilon[N_1(t + \Delta t] - N_1(t)] - \varepsilon[N_2(t + \Delta t) - N_2(t)] \\
&= \varepsilon \Delta N_1 - \varepsilon \Delta N_2.
\end{aligned}$$

One can also write the following stochastic differential equation involving Poisson processes:

$$dX_\varepsilon = \varepsilon[dN_1 - dN_2].$$

We will now determine the first two infinitesimal moments of a diffusion process, X, obtained from the X_ε-processes as the jump magnitudes ε go to zero and the jump rates λ go to infinity - but in such a way that these moments remain finite and non-zero in the limit. We may call this a **diffusion approximation** to the original discontinuous random walk - and this approximation will have, as we have seen, continuous sample paths. The situation is illustrated in Fig. 12.3.

Now as we have seen in Equation (9.2), the increments in the Poisson process N_1 are such that:

$$\Delta N_1 = \begin{cases} 1, & \text{with prob.} \quad \lambda \Delta t + o(\Delta t); \\ 0, & \text{with prob.} \quad 1 - \lambda \Delta t + o(\Delta t), \end{cases}$$

and the probabilities of other values are $o(\Delta t)$; similarly for increments ΔN_2 in the process N_2.

Thus, utilizing the fact that the increments are independent of previous values and that the two Poisson processes are independent, we have

$$E[\Delta X_\varepsilon | X_\varepsilon(t) = x] = o(\Delta t) \tag{12.31}$$

and

$$\text{Var}[\Delta X_\varepsilon | X_\varepsilon(t) = x] = \varepsilon^2[\text{Var}(\Delta N_1) + \text{Var}(\Delta N_2)] \tag{12.32}$$

$$= 2\varepsilon^2 \lambda \Delta t + o(\Delta t).$$

The first two infinitesimal moments, α_ε and β_ε, of X_ε are thus, from (12.31)

Figure 12.3 Showing how the discontinuous random walk is approximated by a diffusion process with continuous sample paths.

and (12.32) respectively, and the definitions (12.1) and (12.2),

$$\alpha_\varepsilon(x, t) = \lim_{\Delta t \to 0} \frac{o(\Delta t)}{\Delta t} = 0,$$

and

$$\beta_\varepsilon(x, t) = \lim_{\Delta t \to 0} \frac{2\varepsilon^2 \lambda \Delta t}{\Delta t} + \frac{o(\Delta t)}{\Delta t}$$

$$= 2\varepsilon^2 \lambda.$$

Let us now make ε vanish and λ grow to infinity but insist that the infinitesimal variance remains finite and nonzero. This can be easily achieved if, for example, we require

$$\lambda = \frac{1}{2\varepsilon^2}. \tag{12.33}$$

In this case, the first and second infinitesimal moments of the limiting process X are

$$\alpha(x, t) = 0,$$

and

$$\beta(x, t) = 1.$$

With our choice of the relation (12.33) and letting $\varepsilon \to 0$, it will be seen in the exercises that all the higher order infinitesimal moments of X vanish as required by (12.3) – although some are zero regardless of the value of ε.

We can now use Theorem 12.5. Since $\beta(x,t) = 1$ we have $g^2(x,t) = 1$, so, choosing the positive root,

$$g(x,t) = 1.$$

Hence the diffusion process approximating the random walk has the stochastic differential equation

$$dX = dW.$$

That is, X is a standard Wiener process. One may also illustrate this **convergence to a Wiener process** using characteristic functions - see the exercises.

A mathematical model for the activity of a nerve cell (neuron)

In Fig. 12.4 we depict a *nerve cell* or *neuron* – a component of the nervous system. The one depicted is a pyramidal cell of the part of the human brain called the cerebral cortex. Such cells are specialized to receive signals from other cells at the junctions called *synapses*. If no signals are received, the electrical potential difference V_M across the target cell membrane tends to a fixed value called the *resting potential*, which we designate by V_R. Usually V_R is about 70 millivolts (the inside being negative) or about $\frac{1}{20}$th of the voltage of the familiar AA battery.

Let us denote the difference between V_M and V_R by V:

$$V = V_M - V_R,$$

- see the left part of Fig. 12.5. V is called the *depolarization*.

As explained at the end of Chapter 7, the incoming signals may *excite* the target cell: this makes V_M less negative, in which case V is increased. We assume that each such incoming signal increases V by an amount a_E and that such signals occur as a Poisson process N_E with intensity λ_E. Alternatively an incoming signal may *inhibit* the target cell, thereby decreasing V (which can go negative), by an amount a_I. Let such inhibitory effects be generated by a Poisson process N_I with intensity λ_I. These events result in V's executing a random walk as depicted on the right in Fig. 12.5.

The state of the neuron at time t can thus be characterized by

$$V(t) = a_E N_E(t) - a_I N_I(t),$$

and we assume $V(0) = 0$, so the cell is initially at resting level. The reader can show that the mean and variance of the change in V during $(t, t + \Delta t]$ are

$$E[\Delta V] = a_E \lambda_E \Delta t - a_I \lambda_I \Delta t + o(\Delta t),$$

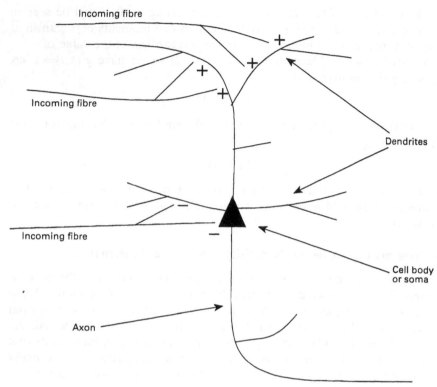

Figure 12.4 Here we depict a class of neuron called a pyramidal cell, showing the main components of axon, dendrites and soma. The neuron is shown connected into a network through synapses at which other cells transmit signals to the target cell. Here a (+) indicates an excitatory junction whereas a (–) denotes an inhibitory one.

and

$$\mathrm{Var}[\Delta V] = a_E^2 \lambda_E \Delta t + a_I^2 \lambda_I \Delta t + o(\Delta t).$$

Dividing by Δt and taking the limit $\Delta t \to 0$, we obtain the first and second infinitesimal moments. We may construct a diffusion approximation $V^*(t)$ to V with the same infinitesimal mean and variance, viz.,

$$\alpha(x, t) = a_E \lambda_E - a_I \lambda_I,$$
$$\beta(x, t) = a_E^2 \lambda_E + a_I^2 \lambda_I.$$

Using Theorem 12.5, we find the following stochastic differential for V^*:

$$dV^* = (a_E \lambda_E - a_I \lambda_I)\,dt + \sqrt{a_E^2 \lambda_E + a_I^2 \lambda_I}\,dW.$$

This will be recognized as the stochastic equation for a *Wiener process with drift*.

Figure 12.5 On the left we show the relation between V, V_R and V_M. On the right are shown the variations which occur in V as signals come in from other cells. Two excitatory inputs (Ex.) and one inhibitory input (Inhib.) are depicted. The solid line is a trajectory for the process $a_E N_E(t) - a_I N_I(t)$ used in the text to model V. The dashed line represents the more physiologically realistic trajectory involving decay to resting level ($V = 0$) between input events.

The mean and variance, as well as the transition density of the state of a nerve cell can be estimated using this approximation. In addition, the time taken for V or V^* to reach a **threshold value**, θ, say, can be determined as a **first passage time**. When V reaches this threshold value, which is usually about 10 millivolts (i.e., 0.01 volts) from resting level, our target cell itself sends out a signal along its axon (see Fig. 12.3) to excite or inhibit other neurons. The whole collection of neurons is sometimes called a **neural network**. Details of such calculations can be found in Tuckwell (1988) where it is also seen that the *Ornstein–Uhlenbeck process* (see section 12.4) is a somewhat more realistic mathematical model for a neuron than the Wiener process with drift. In the absence of inputs the OUP model correctly predicts that the electrical potential across the neuron's membrane will return exponentially to resting level.

A model for population growth in a random environment

As we have seen in section 9.1, when a population grows in an unrestrained way, the population size $x(t)$ at time t may be approximated by the solution of the deterministic differential equation

$$\frac{dx}{dt} = rx, \tag{12.34}$$

where the assumed constant growth rate r is the difference between the birth and death rates. This is the *Malthusian law* with exponential solutions

$$x(t) = x(0) e^{rt},$$

$x(0)$ being the initial population size.

There are often factors, particularly environmental ones, which make the birth and death rates chop and change from generation to generation. For example, in very cold or very dry seasons we might expect a higher death rate in populations of many species, so r would drop below some long-term average value. The same might happen in very hot and very wet seasons. On the other hand, if climatic conditions lead to an abundance of food, r might increase above its average value, but this would be complicated if an organism's predators also benefited.

This suggests that a more accurate mathematical model of population growth would result if the growth rate r and hence the population size x are random processes. The simplest (but not necessarily accurate) assumption is that the growth rate $r(t)$ is the sum of its constant mean value $\bar{r}$ and a fluctuating white noise,

$$r(t) = \bar{r} + \sigma w(t),$$

where σ is a variance parameter and $w(t)$ is standard white noise, defined in section 11.3. Then we may write the model for the random population size $X(t)$,

$$\frac{dX}{dt} = (\bar{r} + \sigma w(t))X(t),$$

or, more satisfactorily as the stochastic differential equation,

$$dX = \bar{r}X\, dt + \sigma X\, dW. \tag{12.35}$$

We let the initial value be

$$X(0) = x_0.$$

As mentioned in the last section, the approach we have used has the apparent drawback of necessitating a choice of stochastic integral. Fortunately, however, in this case the derivation of a diffusion approximation to Malthusian growth has been carried out (Tuckwell and Walsh, 1983a) from first principles. This approximation satisfies Equation (12.35) interpreted as a Stratonovich equation. Note that this does not imply that the Stratonovich integral is better or more correct than the Ito integral.

If we proceed with Equation (12.35) as a Stratonovich equation we can use the usual rules of calculus. In particular, we find that the change of variable

$$Y = \ln(X)$$

leads to

$$\frac{dY}{dt} = \frac{dY}{dX}\frac{dX}{dt}$$

$$= \frac{1}{X}\frac{dX}{dt}.$$

Thus

$$\frac{dY}{dt} = \bar{r} + \sigma\frac{dW}{dt}$$

or, equivalently,

$$dY = \bar{r}\,dt + \sigma\,dW.$$

Thus the transformed process Y is simply a *Wiener process with drift*. This is a particular case of a general transformation method (Tuckwell, 1973; 1974) in which the equation

$$dX = f(X)[\mu\,dt + \sigma\,dW]$$

is transformed to a Wiener process with drift using the transformation

$$Y = \int^{X} \frac{1}{f(X')}\,dX'$$

– see the exercises.

Note that in our present application the possible values of X are in $(0, \infty)$ whereas the possible values of Y are in $(-\infty, \infty)$. In addition, if $X(0) = x_0$ and $Y(0) = y_0$, then $y_0 = \ln(x_0)$.

The transition density of a Wiener process with drift was given by Equation (12.15). If, in conjunction with that result, we use Equation (1.6) for the transformation of a density under a monotonic change of variable, we are immediately able to find the **transition probability density of the population size X** as

$$p(x, t\,|\,x_0) = \frac{1}{x\sqrt{2\pi\sigma^2 t}}\exp\left(\frac{-[\ln(x/x_0) - \bar{r}t]^2}{2\sigma^2 t}\right).$$

Now, we can see that the **probability P_E that the population eventually becomes extinct** can be estimated from

$$P_E = \lim_{t \to \infty} \Pr\{X(t) < \varepsilon\,|\,X(0) = x_0\}$$

where $\varepsilon > 0$ is arbitrarily small. Note that X can never reach exactly zero in this model – for we know that Y can never reach $-\infty$. However, an extremely small population size implies extinction in a continuous model.

It will be seen in the exercises that

$$P_E = \begin{cases} 0, & \text{if } \bar{r} > 0; \\ \frac{1}{2}, & \text{if } \bar{r} = 0; \\ 1, & \text{if } \bar{r} < 0. \end{cases}$$

Thus, if the mean growth rate is negative, the population will become extinct with certainty, regardless of either the initial population size or how great the fluctuations (σ) are in the growth rate. In this model, if a zero population growth policy is sustained on average, there is a 50% chance of long-term survival.

In addition, diffusion processes and stochastic differential equations have been usefully employed to model gene frequencies (Kimura, 1964; Tuckwell, 1976a; Watterson, 1979) – providing continuous approximations which are usually easier to work with than the Markov chain models we considered in Chapter 8, where additional references can be found. With these tools, important questions in the theory of evolution can be addressed.

Applications in financial modelling

As exemplified by Figure 11.2b, many quantities of financial or economic interest undergo random fluctuations. Needless to say, if the random component was absent, stock prices, for example, would be predictable so that there would be no risk involved in buying or selling shares. This would imply little possibility of profit and stock markets might cease to exist *per se*.

It is clear that a sound mathematical model for fluctuations in financial entities such as stock and commodity prices, exchange rates, etc. would be very useful as one could then make quantitative estimates for the probabilities of future values, expected profits, waiting times to reach certain levels, etc. .

As early as 1900, the French mathematician Bachelier proposed the modelling of stock market fluctuations using a Wiener process. It was subsequently recognized that because the prices of shares could not be negative, a modification to the simple Wiener process was necessary. One solution proposed was the adoption of so-called *geometric Brownian motion* or the *geometric Wiener process* to represent certain financial entities. Thus $Y = \ln(X)$ should be a Wiener process with drift – so that $X = e^Y$ might represent a stock price. We then have $Y \in (-\infty, \infty)$ but $X \in (0, \infty)$.

Another assumption can be made that the expectation of a stock price grows in an exponential fashion (Samuelson, 1965). Thus if $X(t)$ is a stock price at time t, then

$$E[X(t)|X(0) = x_0] = x_0 e^{\mu t}, \tag{12.36}$$

where μ represents a constant growth rate.

To satisfy these requirements we need only put

$$\mathrm{d}X = [\mu\,\mathrm{d}t + \sigma\,\mathrm{d}W]X, \tag{12.37}$$

and interpret this as in Ito stochastic differential equation. That this gives the correct relation (12.36) will be verified in the exercises. The process X is called geometric Brownian motion and was analysed in the context of general growth models, and in particular tumour growth, by Smith and Tuckwell (1974).

In the economics and finance literature, Equation (12.37) is often written

$$\frac{\mathrm{d}X}{X} = \mu\,\mathrm{d}t + \sigma\,\mathrm{d}W.$$

This is an essential component of the Black–Scholes model for option prices (Black and Scholes, 1973) which is often used by financial analysts. This model, which provides a starting point for more elaborate models (Aase, 1988), leads to a formula called the *Black–Scholes option formula* for an option price in terms of variables such as term to maturity and risk-free interest rate. Wiener process models have been employed in the analysis of exchange rates also (Werner, 1993).

Practical considerations – simulation, numerical methods and parameter estimation

When employing stochastic differential equations as mathematical models of empirical phenomena, it is often worthwhile, and in many cases necessary, to perform computer simulations. The computer simulation of diffusion processes is not difficult (Tuckwell and Walsh, 1983b) and can be performed as follows.

Let us suppose that we wish to approximate solutions of the Ito equation

$$\mathrm{d}X = f(X,t)\,\mathrm{d}t + g(X,t)\,\mathrm{d}W.$$

We choose a time step Δt and call our simulation X^* with

$$X^*(k\Delta t) \simeq X(k\Delta t)$$

where $k = 0, 1, 2, \dots$. We are given (or draw from a distribution) the value $X^*(0) = X(0)$. Then we put, for $k = 1, 2, \dots$

$$X^*(k\Delta t) = X^*((k-1)\Delta t) + f(X^*(k-1)\Delta t, (k-1)\Delta t))\Delta t$$
$$+ g(X^*((k-1)\Delta t), (k-1)\Delta t)\sqrt{\Delta t}\,N_k, \tag{12.38}$$

where $\{N_k, k = 1, 2, \dots\}$ is a sequence of i.i.d. standard normal random variables. The latter can be generated by the methods outlined in Chapter 5 – or simply by using a library random number generator. One performs a large number of runs or trials for the process and can easily determine estimates of quantities such as moments, distributions, and first passage or first exit times, etc. .

An alternative to (12.38) is to employ a sequence of independent Bernoulli random variables B_k (see page 2) instead of the N_k with

$$\Pr\{B_k = 1\} = 1/2 = \Pr\{B_k = -1\}, k = 1, 2, \ldots .$$

Then since $E[B_k] = 0$ and $\mathrm{Var}[B_k] = 1$, one may simply replace N_k in (12.38) by B_k. Since most computers have in their libraries a generator of uniformly distributed random variables (see Chapter 5), U_k, say, the sequence of Bernoulli variables can be easily obtained by putting $B_k = 1$ if $U_k < \frac{1}{2}$ or $B_k = -1$ if $U_k \geqslant \frac{1}{2}$. It has been found that when Bernoulli rather than normal random variables are employed, the simulations may be performed about three times faster (Tuckwell and Lansky, 1994). A detailed treatment of theoretical considerations on the numerical solutions of stochastic differential equations can be found in Kloeden and Platen (1992).

As mentioned earlier, another approach to the study of diffusion processes is to use the analytical method. This will usually involve solving a Kolmogorov or similar equation. Fortunately, even though the stochastic differential equation may be nonlinear, the Kolmogorov equation is always linear. It is then simply a matter of solving the latter using finite-difference approximations either by explicit or implicit methods (Ames, 1977). The analytical approach to first passage and first exit time problems is outlined in Siegert (1951), Tuckwell (1976b; 1981) and Tuckwell and Wan (1984).

Finally, assuming that one has a satisfactory mathematical representation, in the form of a stochastic differential equation, for an empirical process that one wishes to model, it is often desirable to estimate values of the various parameters which occur in the model using observed sample paths. Various methods are available, including maximum likelihood (Feigin, 1976) and quasi-likelihood (Heyde, 1993).

REFERENCES

Aase, K.K. (1988). Contingent claims valuation when the security price is a combination of an Ito process and a random point process. *Stoch. Proc. Appl.*, **28**, 185–220.
Ames, W.F. (1977). *Numerical Methods for Partial Differential Equations*. Nelson, London.
Arnold, L. (1974). *Stochastic Differential Equations*. Wiley, New York.
Bachelier, L. (1900). Theorie de la speculation. *Ann. Sci. Ecole. Norm. Sup.*, **3**, 21–86. English translation in Cootner, P. H. (ed.) (1964), *The Random Character of Stock Market Prices*, pp. 17–78. MIT Press, Cambridge.
Berkey, D.D. (1984). *Calculus*. Holt-Saunders, Philadelphia.
Black, F. and Scholes, M. (1973). The pricing of options and corporate liabilities. *J. Pol. Econ.*, **81**, 637–54.

Chandrasekhar, S. (1943). Dynamical friction. *Astrophys. J.*, **97**, 263–73.

Feigin, P.D. (1976). Maximum likelihood estimation for continuous-time stochastic processes. *Adv. Appl. Prob.*, **8**, 712–36.

Feller, W. (1952). The parabolic differential equations and the associated semi-groups of transformations. *Ann. Math.*, **55**, 468–519.

Feller, W. (1971). *An Introduction to Probability Theory and its Applications II.* Wiley, New York.

Gihman, I.I. and Skorohod, A.V. (1972). *Stochastic Differential Equations.* Springer-Verlag, Berlin.

Gradshteyn, I.S. and Ryzhik, I.M. (1980). *Tables of Integrals Series and Products.* Academic Press, New York.

Heyde, C.C. (1993). Quasi-likelihood and general theory of inference for stochastic processes. *Lecture Notes. 7th International Summer School on Probability Theory and Mathematical Statistics* (eds A. Obretenov and V.T. Stefanov), pp. 122–52. Science Culture Technology Publishing, Singapore.

Hida, T. (1980). *Brownian Motion.* Springer-Verlag, Berlin.

Jaswinski, A.H. (1970). *Stochastic Process and Filtering Theory.* Academic Press, New York.

Kimura, M. (1964). Diffusion models in population genetics. *J. Appl. Prob.*, **1**, 177–232.

Kloeden, P. and Platen, E. (1992). *The Numerical Solution of Stochastic Differential Equations.* Springer-Verlag, Berlin.

Liptser, R.S. and Shiryayev, A.N. (1977). *Statistics of Random Processes I. General Theory.* Springer-Verlag, New York.

McNeil, D.R. and Schach, S. (1973). Central limit analogues for Markov population processes. *J. Roy. Stat. Soc. B.*, **35**, 1–23.

Noble, B. and Daniel, J.W. (1977). *Applied Linear Algebra.* Prentice-Hall, Englewood Cliffs.

Oksendal, B. (1985). *Stochastic Differential Equations.* Springer-Verlag, Berlin.

Ricciardi, L.M. (1977). *Diffusion Processes and Related Topics in Biology.* Springer-Verlag, Berlin.

Samuelson, P.A. (1965). Rational theory of warrant pricing. *Indust. Manag. Rev.*, **6**, 13–31.

Shields, P.C. (1973). *Elementary Linear Algebra.* Worth, New York.

Siegert, A.J.F. (1951) On the first passage time probability problem. *Phys. Rev.* **81**, 617–23.

Smith, C.E. and Tuckwell, H.C. (1974). Some stochastic growth processes. In *Mathematical Problems in Biology Proceedings 1973*, pp. 211–25. Springer-Verlag, Berlin.

Stratonovich, R.L. (1963). *Topics in the Theory of Random Noise.* Gordon and Breach, New York.

Tuckwell, H.C. (1973). Diffusion Processes in Biology. *SM Thesis.* University of Chicago.

Tuckwell, H.C. (1974). A study of some diffusion models of population growth. *Theor. Pop. Biol.*, **5**, 345–57.

Tuckwell, H.C. (1976a). The effects of random selection on gene frequency. *Math. Biosci.*, **30**, 113–28.

Tuckwell, H.C. (1976b). On the first exit-time problem for temporally homogeneous Markov processes. *J. Appl. Prob.*, **13**, 39–48.

Tuckwell, H.C. (1981). Poisson processes in biology. In *Stochastic Nonlinear Systems.* pp. 162–71, Springer-Verlag, Berlin.

Tuckwell, H.C. (1987). Diffusion approximations to channel noise. *J. Theor. Biol.*, **127**, 427–38.

Tuckwell, H.C. (1988). *Introduction to Theoretical Neurobiology, Vol. 2, Nonlinear and Stochastic Theories.* Cambridge University Press, New York.

Tuckwell, H.C. (1994). *Stochastic Processes in the Neurosciences.* 2nd edn. SIAM, Philadelphia.

Tuckwell, H.C. and Lansky, P. (1994). On simulation methods for stochastic differential equations. *Stochastic Analysis Group Research Report, Australian National University.*

Tuckwell, H.C. and Walsh, J.B. (1983a). Stochastic calculus and diffusion models in population biology. *Institute of Applied Mathematics and Statistics, University of British Columbia,* Technical Report **59**.

Tuckwell, H.C. and Walsh, J.B. (1983b). Random currents through nerve membranes. *Biol. Cybern.* **49**, 99–110.

Tuckwell, H.C. and Wan, F.Y.M. (1984). First passage time of Markov processes to moving barriers. *J. Appl. Prob.* **21**, 695–709.

Uhlenbeck, G.E. and Ornstein, L.S. (1930). On the theory of Brownian motion. *Phys. Rev.,* **36**, 823–41.

Watterson, G.A. (1979). Estimating and testing selection: the two-alleles, genic selection diffusion model. *Adv. Appl. Prob.,* **11**, 14–30.

Werner, A.M. (1993). Exchange rates and target zone width. *Econ. Lett.,* **40**, 455–7.

Wong, E. and Zakai, M. (1965). On the convergence of ordinary integrals to stochastic integrals. *Ann. Math. Stat.,* **36**, 1560–64.

EXERCISES

Note: Problems 4, 9, 22, 23 and 29 may be suitable for project work. In addition the instructor may favour the setting of project work which extends the models in biology and finance of section 12.7 or considers similar models in other areas.

1. Show that the second infinitesimal moment, defined in Equation (12.2), is also given by

$$\beta(x,t) = \lim_{\Delta t \to 0} \frac{\mathrm{Var}[\Delta X \mid X(t) = x]}{\Delta t}.$$

2. Verify that for a time-homogeneous diffusion process the Kolmogorov equations become Equations (12.6) and (12.7).

3. For the unrestricted Wiener process, sketch what happens to the transition density as $t \to \infty$ for the cases $\mu < 0$, $\mu = 0$ and $\mu > 0$.

4. Outline how you would apply the method of separation of variables to find the transition density for a Wiener process with drift on $(0, a)$ with absorbing barriers at 0 and a.

5. Show by integration with the transition density (12.15) of a Wiener process with drift, that the mean and variance are $x + \mu t$ and $\sigma^2 t$, respectively.

6. Show that the spatial eigenfunctions for the Wiener process between reflecting barriers satisfy the orthogonality relation (12.17) and determine the constants A_n. (*Hint:* A handbook of integrals such as Gradshteyn and Ryzhik (1980) may be useful.)

7. Verify Equation (12.19) and use it to investigate the behaviour of $p_A(t; a, x)$ as $t \to \infty$ for various values of μ.

8. Use the theory of section 12.2 to verify that the invariant distribution in the case of a Wiener process confined between reflecting barriers is as given in section 12.3. Sketch the density for the following parameter values: $a = 2$, $\sigma^2 = 1$, $\mu = -1$, 0 and 1.

9. Repeat Exercise 4 for the OUP with reflecting barriers at 0 and a.

10. Use the formulae of section 12.2 to verify that the stationary density for an unrestricted OUP is as given in section 12.4.

11. Write down a stochastic differential equation (SDE) for an OUP with infinitesimal drift and diffusion given by $\alpha(x, t) = -ax + b$ and $\beta(x, t) = \sigma^2$, respectively. What would be the difference between an Ito or Stratonovich interpretation of this SDE? Take expectations in the SDE and integrate to find the mean of the process at time t.

12. Show that the expected value of the Ito integral of a simple random process is zero.

13. Let X be a standard Wiener process. Use Ito's formula to obtain Ito SDEs satisfied by $Y = $ (a) $X + t^2$, (b) $X^2 + t^2$ and (c) $\ln X$.

14. Convert the following Stratonovich SDEs to Ito equations: (a) $dX = X \, dt + \sqrt{X} \, dW$; (b) $dX = X^2 \, dt + e^{-t} \, dW$; (c) $dX = X^3 \, dt + e^{-t} X \, dW$.

15. The deterministic logistic growth equation $dx/dt = x(K - x)$, where $K > 0$ is the carrying capacity, can be converted to a model for population growth in a random environment by replacing K with the randomly fluctuating quantity $\bar{K} + \sigma w(t)$, where $\bar{K}$ is the average carrying capacity and w is a white noise. Demonstrate, by considering invariant distributions, that critical differences may occur between the use of an Ito or Stratonovich interpretation.

16. Show that in the limit $\varepsilon \to 0$ the infinitesimal moments of order greater than two vanish in the sequence of processes used to model the activity of a nerve cell, $dX_\varepsilon = \varepsilon[dN_1 - dN_2]$.

17. Use characteristic functions to prove the convergence of the random walk model of section 12.7 to a Wiener process.

18. Verify the expressions in section 12.7 for the mean and variance of ΔV.

19. Show that the Stratonovich SDE $dX = f(X)[\mu \, dt + \sigma \, dW]$ may be transformed to that of a Wiener process by setting

$$Y = \int^X \frac{1}{f(X')} \, dX',$$

so that the transition density of X may be obtained as

$$p(x, t \,|\, x_0) = \frac{|f(x)|^{-1}}{\sqrt{2\pi\sigma^2 t}} \exp\left[-\frac{\left(\int_{x_0}^x \frac{1}{f(x')} \, dx' - \mu t \right)^2}{2\sigma^2 t} \right].$$

20. Show that in the population model, Equation (12.35), the probability of extinction P_E is as given in the text.

21. Show that in the stock price model of section 12.7, the mean price is $x_0 e^{\mu t}$.

22. Using the method outlined in section 12.7, perform a simulation of the process defined by the stochastic differential equation $dX = X[\mu \, dt + \sigma \, dW]$ using both Ito and Stratonovich interpretations. Vary the number of trials and the time-step Δt, for various parameter values and values of $X(0)$. Find estimates for the means and variances as functions of time (confidence intervals can also be determined) and compare these with the values obtained using the exact expressions. Determine also empirical distributions (histograms) for the time taken for the process to first attain certain values.

23. Perform a simulation for the OUP with your own choice of parameter values and determine the same properties as in Exercise 22. In addition, compare the density approached by your simulated process for large times with the exact expression for the stationary density in section 12.4.

24. Compare the properties of the birth and death process of sections 9.7 and 9.8 with those of the random environment growth process of section 12.7.

25. Show that the spatial eigenfunctions of the Kolmogorov equation on finite spatial intervals are orthogonal with weight factor p^*.

26. Determine the trajectories followed by an OUP when $\sigma = 0$. Do these coincide with $E[X(t)| X(0) = x]$?

27. Let an OUP satisfy the SDE $dX = (\mu - X) dt + \sigma \, dW$, with $X(0) = x$. Show that the transformed process $Y = e^t X$ satisfies $dY = e^t(\mu \, dt + \sigma \, dW)$. Hence

$$Y(t) = Y(0) + \mu \int_0^t e^{t'} \, dt' + \sigma \int_0^t e^{t'} \, dW(t').$$

Now determine the mean and variance of $Y(t)$ using (12.24) and (12.25). Hence verify the formulae for the mean and variance of the OUP given in the text and the expression in section 12.4 for the transition density of an OUP.

28. Let X be an OUP satisfying $dX = (\mu - \alpha X) dt + \sigma \, dW$. Find $\text{Cov}[X(s), X(t)]$.

29. Show that infinite boundary points are *natural* and finite boundary points are *regular* for the OUP.

30. Using a computer, solve the forward Kolmogorov equation for the Wiener process with drift using your own choice of parameters and a numerical method of solution such as that of Crank-Nicolson. Use an initial density which is Gaussian and compare your computed solutions with the exact solution.

Appendix

Table of critical values of the χ^2-distribution (see Section 1.7), v = degrees of freedom

v	α							
	0.995	0.99	0.975	0.95	0.05	0.025	0.01	0.005
1	0.0^4393	0.0^3157	0.0^3982	0.0^2393	3.841	5.024	6.635	7.879
2	0.0100	0.0201	0.0506	0.103	5.991	7.378	9.210	10.597
3	0.0717	0.115	0.216	0.352	7.815	9.348	11.345	12.838
4	0.207	0.297	0.484	0.711	9.488	11.143	13.277	14.860
5	0.412	0.554	0.831	1.145	11.070	12.832	15.086	16.750
6	0.676	0.872	1.237	1.635	12.592	14.449	16.812	18.548
7	0.989	1.239	1.690	2.167	14.067	16.013	18.475	20.278
8	1.344	1.646	2.180	2.733	15.507	17.535	20.090	21.955
9	1.735	2.088	2.700	3.325	16.919	19.023	21.666	23.589
10	2.156	2.558	3.247	3.940	18.307	20.483	23.209	25.188
11	2.603	3.053	3.816	4.575	19.675	21.920	24.725	26.757
12	3.074	3.571	4.404	5.226	21.026	23.337	26.217	28.300
13	3.565	4.107	5.009	5.892	22.362	24.736	27.688	29.819
14	4.075	4.660	5.629	6.571	23.685	26.119	29.141	31.319
15	4.601	5.229	6.262	7.261	24.996	27.488	30.578	32.801
16	5.142	5.812	6.908	7.962	26.296	28.845	32.000	34.267
17	5.697	6.408	7.564	8.672	27.587	30.191	33.409	35.718
18	6.265	7.015	8.231	9.390	28.869	31.526	34.805	37.156
19	6.844	7.633	8.907	10.117	30.144	32.852	36.191	38.582
20	7.434	8.260	9.591	10.851	31.410	34.170	37.566	39.997
21	8.034	8.897	10.283	11.591	32.671	35.479	38.932	41.401
22	8.643	9.542	10.982	12.338	33.924	36.781	40.289	42.796
23	9.260	10.196	11.689	13.091	35.172	38.076	41.638	44.181
24	9.886	10.856	12.401	13.848	36.415	39.364	42.980	45.558
25	10.520	11.524	13.120	14.611	37.652	40.646	44.314	46.928
26	11.160	12.198	13.844	15.379	38.885	41.923	45.642	48.290
27	11.808	12.879	14.573	16.151	40.113	43.194	46.963	49.645
28	12.461	13.565	15.308	16.928	41.337	44.461	48.278	50.993
29	13.121	14.256	16.047	17.708	42.557	45.722	49.588	52.336
30	13.787	14.953	16.791	18.493	43.773	46.979	50.892	53.672

Reproduced with the permission of *Biometrika*, from *Tables for Statisticians*. In the first row 0^4, for example, means four zeros.

Index

Page numbers appearing in *italic* refer to figures.

Milton Keynes UK
Ingram Content Group UK Ltd.
UKHW021618071024
449327UK00020BA/1099